A STUDY OF RADNORSHIRE PLACE-NAMES

A study of
RADNORSHIRE PLACE-NAMES

Richard Morgan

GWASG Carreg Gwalch

ISBN: 0-86381-487-5

Cover design: Smala

First published in 1998 by Gwasg Carreg Gwalch,
12 Iard yr Orsaf, Llanrwst, Wales LL26 0EH
☎ (01492) 642031
Printed and published in Wales.

Contents

Introduction

I have been interested in the subject of place-names for about twenty years, initially as a sideline to other historical matters. Anyone who has undertaken research in medieval history will be aware how difficult and important it can be in identifying places recorded in documents, for example. When I lived in England I could consult Ekwall's DEPN, often the relevant county volume(s) of the EPNS and the Victoria County History, as well as gazetteers and directories in identifying place-names correctly. If I was interested in the meaning of a particular place-name I could rely on Ekwall, the EPNS or the EPNE.

When I returned to Wales it was quite different: there were and are no general surveys of place-name elements apart from that of Mawer, no dictionaries to match the DEPN, no reliable surveys of place-name elements and no societies specialising in the study of place-names. Even our authoritative dictionary GPC is incomplete and has not been able to rely upon comprehensive place-name research. There are, of course, some excellent local studies[1] and some surveys on particular themes[2] but the blunt fact remains that for anyone with an interest in the subject of Welsh place-names it can be quite disheartening. If you want answers to your questions then you will be forced very often to rely upon unreliable antiquarians, outdated local histories and guides, and dictionaries whose primary function is to define *words* not place-names. So often those who take the subject seriously find themselves working in near-isolation, scratching around for authoritative and comparative evidence.

This book arose partly from my dissatisfaction with the state of Welsh place-name studies and partly from my belief that I could make a positive contribution: a popular study of a county's place-names which would be of use to other local historians and to anyone with an interest in the general subject

---

[1] Notably PNDP; PNPEM; PNEF and see Bibliography below.

[2] e.g. EANC and NCPN

of place-names in Wales and indeed the rest of Britain. What follows represents only a small part of the work which I have compiled; the fuller survey includes the place-names of Montgomeryshire and Breconshire and many minor place-names in Radnorshire.

On one matter I am quite certain: this is not meant to be the definitive work on the place-names of Radnorshire, partly because it omits sources, partly because I am not a place-name specialist and partly because definitive studies require more extensive analysis of place-names and minor names such as those of fields and rivers. There has been no research on these subjects so far as I am aware in this county. I am conscious too that for many of the place-names the selection of spellings is either too sparse or too late to sustain reliable interpretations and analysis. No doubt some of my interpretations will have to be revised as historical and place-name study advances.

There have been plenty of complaints about the backward state of Welsh place-name studies[3] and a lot of advice on the best method of tackling the subject [4] yet publications by what Richards called 'the not so gifted amateur' continue to appear and many a *modern* parish history or church guide contains silly place-name interpretations. Part of the reason for this state-of-affairs may lie in a supposition that Welsh place-names have little to tell us by contrast with English place-names and that there is little point in studying the obvious. After all, most Welsh place-names do not appear to be meaningless or obscure to Welsh speakers: you do not need a PhD in Welsh to render the meaning of Bwlchysarnau, Esgairnantau or Llangynllo.

The trouble about this supposition is that it relies too heavily upon modern spellings and modern dictionaries: it is easy to make mistakes since modern spellings can be misleading; note,

---

3 S.J. Williams, *Brycheiniog* 10, 155; Melville Richards in *Mont.Colls.* 54, 97.

4 D.M. Ellis, *Mont.Colls.* 44, 87-88; G.O. Pierce, 'Place-Names', *Settlement and Society in Wales*, ed. D. Huw Owen (Cardiff 1984)

for example, Perwydd, the map form *Cwmdauddwr* (see Cwmteuddwr), Nantmel and Caerfaelog; it would be very easy to misinterpret a name such as Meistirrhosllawddu.

Place-name study amounts to a good deal more than simply rendering a place-name in modern language or translating from one language to another. Analysing past spellings can help us in the task of correctly identifying places mentioned in historic texts and records (see Aberedw for an example) and can help us in the study of the history of language and dialect – particularly important in a county like Radnorshire which has lost its Welsh and its distinctive dialect. Too often the debate has centred on the 'correctness' of a particular spelling.

The unwillingness of some localities to adopt the standard Welsh spelling (e.g. Llaethdy) tends to attract the attention of the public media. The reluctance of some towns and villages to alter the spelling of their place is understandable but their arguments are sometimes based on the assumption that the modern spelling has 'always' been the one used. It is easily forgotten that the modern spelling may be very recent in historical terms. The treatment of place-names has differed too according to whether a particular place-name has been perceived as Welsh or English. Over the past 200 years Welsh place-names have been gradually standardised: *Llanelly* gave way to *Llanelli, Carnarvon* to *Caernarfon,* but English place-names have resisted such changes, largely because the spelling of English in general has resisted 'rationalisation'.

MAP 1

RADNORSHIRE Commotes and Cantrefi

The Welsh Language in Radnorshire

Most place-names in this book are Welsh, a Celtic language closely related to the extinct Welsh language of Strathclyde and Cumbria, and to Cornish and Breton. Welsh, in common with its sister languages, has borrowed hundreds of words from other languages with which it has come into contact, particularly Latin and English. It has also suffered recession and decline, though not as disastrously as Cornish, Manx and Scots Gaelic. The first statistics which we possess for the number of Welsh speakers are recorded in the 1891 census but they are reckoned to be unreliable.[5] At that time, roughly two-thirds of the people of Wales were able to speak Welsh. Today (1997) the figure is about one-fifth.

Welsh was once spoken over the whole of Wales together with adjoining parts of Shropshire and Herefordshire and small parts of Gloucestershire and Cheshire. Until recently Welsh scholars have shown little interest in describing the history and the historical geography of the languages of Wales – they have included Irish as well as English and Welsh – and dating the ebb and flow of the languages in Wales. Thankfully, the position has now changed and Wales is now gaining a first-class study of the Welsh language. [6] It is a difficult subject, of course, because 'hard' evidence is scarce before 1891: statistics are usually unreliable and localised.

A fair impression of the relative distribution of Welsh and English in Wales c.1750 can be gained from the map compiled by W.T.R. Pryce[7] based on evidence of the languages used in Anglican churches. In Radnorshire roughly one-sixth of the county used only Welsh or mainly Welsh in church services, compared with about half of Montgomeryshire and

---

5 John E. Southall, *Wales and her Language* (2nd ed., London 1893) 356.

6 See in particular WLIR; Southall, *passim*; A.J. Ellis, 'On the Delimitation of the English and Welsh Languages', *Y Cymmrodor* 5, 173-208.

7 Geraint H. Jenkins, *Wales 1642-1780* (Oxford and Cardiff 1987), 398.

Breconshire. All of the 'Welsh' churches in Radnorshire lay west of a line drawn from Llangurig MTG through Saint Harmon, along the river Wye to a point below Builth and then to Brecon BRE. The 'English only' area already included most of the county. The evidence is plainly incomplete and it can be fairly argued that the language used in church might owe much to the personal preferences of the clergy. There was some antipathy towards the Welsh language in the Church of England. There was certainly a tendency to adopt English services even when only a handful of the population (so far as we can judge from other evidence) spoke the language: the language of the church may not have been the language of the home (or indeed the near-by chapel). Yet these criticisms apply equally well to other parts of Wales. The impression given by Pryce's map is supported by other evidence.

We know that as early as 1684 the parish of Heyope was said to have no Welsh Book of Common Prayer and had had none within the memory of any of the parishioners; it was said that they 'doe all understand English'.[8] Whether they were bilingual or monoglot English speakers is unclear.

If we are to believe Malkin 1807 and Samuel Lewis 1843[9] English dominated most of the county in the early and mid 19th century yet so far as we can judge Welsh was widely spoken, particularly in the western half of the county in the late 18th century. Even on the English border there is evidence for the use of Welsh: James Beaumont (died 1750) of the Gore (Old Radnor), a correspondent and associate of Howel Harris, was said to have been more happy when using Welsh. We have the evidence too of the traveller who made his way from Bridgnorth SHR to Llandrindod in 1743 and was unable to make himself understood to the natives. He mentions Welsh

---

8 G.E. Evans, *Archaeologia Cambrensis* 1918, 110.

9 Benjamin Heath Malkin, *The Scenery, Antiquities, and Biography of South Wales, I* (London 1807), 418-9; Samuel Lewis, *Topographical Dictionary of Wales* (London 1843)

services at Cae-bach chapel (Llandrindod) and Llanbadarn Fawr church, and that the people of Llandrindod spoke only Welsh; Welsh too was the language of Builth market and Aberedw's wakes.[10] Some evidence from other sources is suspect: Lewis Morris says that only Welsh was heard at New Radnor (and more acceptably at Llanfihangel Rhydieithon) in 1742.

There is little doubt that English was spreading in Radnorshire in the 18th century. The fact that Howel Harris held bilingual meetings at Caebanadl (Llanfihangel Nant Melan) in 1745[11] is significant and Lewis Morris himself states that at Llandrindod in 1760 'The Country people talk English . . . as well as Welsh'.[12] By 1827 Llandrindod's young people spoke only English.[13]

The Ecclesiastical Visitation returns of St David's diocese make interesting reading in this respect.[14] In 1733 six parishes, viz. Bugeildy, Heyope, Llangynllo, Pilleth, Whitton and Colfa (and Ceri in the adjoining part of Montgomeryshire) had no Welsh Bibles or Books of Common Prayer. By 1745 Llanddewi Ystradenni and Llandrindod had joined the group. Many parishes are described as 'English cures' or made no use of their Welsh books, e.g. Cleiro and Bugeildy in 1733. Only Saint Harmon was a Welsh cure. That does not mean that Welsh was not still widely spoken – in 1742 the parishioners of Glasgwm petitioned the bishop to send them a vicar who could speak Welsh[15] – but it does suggest that the tide was turning, that Welsh was retreating.

---

[10] Southall, 339-40; *Transactions of the Radnorshire Society* 4, 11-12.

[11] CrF I, 51, 56.

[12] He adds that 'Welsh is more natural to them, and is as pure and correct as in North Wales, some local words excepted': CrF I, 133.

[13] *Gwyliedydd*, Sep. 1837, cited in CrF I, 133-4

[14] *Trans. Rads. Society* 8, 59-76.

[15] *ibid*. 21, 21.

By 1833 English was almost universal in Radnorshire and Welsh was then confined to the area around Rhaeadr, Cwmteuddwr and Saint Harmon.[16] The progress of English is recorded in miscellaneous sources: Abbeycwmhir was reputed to be bilingual c.1810[17]; about the same time Glasgwm had only Christmas morning services in Welsh; and in June 1821 a Baptist *cymanfa* at Newbridge was preached in English as well as Welsh.[18] Weekly Welsh services at Saint Harmon continued till about 1835 and monthly till 1865. In the same area Welsh services survived at Nantgwyn Baptist Chapel till about 1860, Saint Harmon Wesleyan Methodist Chapel till 1870 and Sychnant Calvinistic Methodist Chapel till 1876. In the south of the county Welsh survived only in patches, till about 1850 in Painscastle and about 1890 in Boughrood.[19] Kilvert recorded the death of the last native Welsh speaker in Cleiro in 1867[20] – it reminds one of the last Tasmanian!

Anglicisation had effects on the county's place-names. The most obvious sign was the appearance of English place-names in areas well to the west of the English border such as Newbridge (probably a translation), Black Mountain and Crossgates. The pronunciation of many more place-names changed, some shown in anglicised spellings such as *Goosetree* for Gwystre, others such as 'Lanjewy' (Llanddewi Ystradenni) and 'Cregraina' (Cregrina) show few traces in spelling. In some cases, 'misspellings' were taken on by the OS and came to be regarded as the English forms: note *Rhayader* (Rhaeadr), *Llangunllo* (Llangynllo) and *Beguildy* (Bugeildy). Had it not been

---

16 Lewis, *Top Dict.; Trans.Rads. Society* 22, 76-77.

17 Southall, 341

18 CrF I, 126

19 *Trans.Rads. Society* 4, 35; CrF I, 49. Welsh and English services alternated in Nantgwyllt chapel and both were used in joint services at Saint Harmon, Cwmteuddwr and Llanwrthwl in 1851: *The Religious Census of 1861. A Calendar of the Returns Relating to Wales*, ed. Ieuan Gwynedd Jones and David Williams (Cardiff 1976), I, 662-5.

20 *Kilvert's Diary*, ed. William Plomer (3 vols).

for anglicisation these names would probably have settled down in their standard Welsh spelling. The effects are better represented in house-names and topographical names not in this list; Payne has instanced 'Grappledee' (Craig Pwll Du) and 'Penisha-plwydd' (Penisa'r-plwyf)[21] – though the last is likely to be a dialect sp. – as mispronunciations and gives examples of apparent translations and mistranslations.[22] In some cases the ubiquitous *Llan-* (usually pronounced 'Lan-' here) has submerged *llys, llyn* and *llwyn,* and even *nant.*[23] The pattern is replicated in certain other parts of Wales.[24]

Radnorshire's Place-names

The structure of Welsh place-names and English place-names is quite different and in general it is not a difficult task to determine whether a particular place-name is Welsh or English. In Welsh a compound place-name might consist of a generic and a qualifier, in that order, e.g. Bryn-gwyn, literally 'hill white', and Nant-glas, 'stream blue', sometimes with an intervening definite article, e.g. Maesyronnen, 'field, the ash-tree', where the definite article is prefixed to the qualifier. Recognising this pattern enables us to say that Cwmheyope is essentially a Welsh place-name not an English one despite the fact that *-heyope* is an English element. By contrast we can also recognise Casgob and Disgoed are English place-names despite their Welsh appearance. There are some exceptions to the Welsh pattern, notably with the adjectives *hir* 'long' and *hen* 'old' as in Hirddywel and Hen-goed but there are other examples, e.g. Glasgwm, Gwystre and Marteg. The qualifier usually comes

---

21 CrF I, 102.

22 CrF I, 45; II, 39, 43-44, 82.

23 e.g. *Llys Ifor* has become *Llanshiver*: CrF I, 83. W.H. Howse, *Radnorshire* (Hereford 1949) 301-7; CrF I, 50. *Lloynshaye* 1632 (*Trans.Rads. Society* 5, 55) is now *Llanshay*.

24 See G.O. Pierce, 'The Evidence of Place-Names', *Glamorgan County History II* (Cardiff 1984), 456-92, especially 484-5.

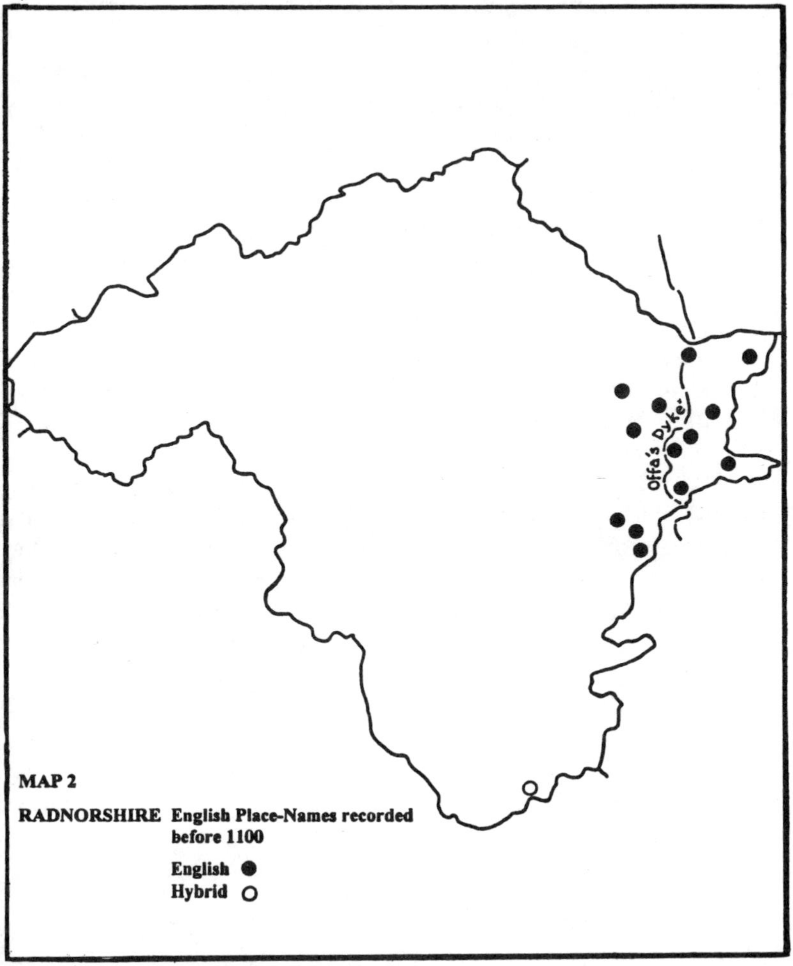

MAP 2

RADNORSHIRE English Place-Names recorded before 1100

English ●
Hybrid ○

first too before the element *ma(n)* 'place' as in Colfa and Cabalfa.

It is important that English readers should bear in mind also the characteristics of the Welsh system of initial mutation – found also in other Celtic languages such as Cornish, Breton and Irish. The most frequent in Welsh place-names is the soft mutation in which initial *b, c, d, ll, p, rh,* and *t* become *f, g, dd, l, b, r* and *d* respectively and in which *g* disappear. As far as place-names are concerned this occurs (a) after an adjective, as in Hen-goed (*coed* to *goed*), (b) after a feminine noun, e.g. Rhos-goch (*coch* to *goch*), (c) to a feminine noun after the definite article, e.g. Pen-y-bont (*pont* to *bont*) and (d) after certain prepositions such as *o* 'from' and *i* 'to'. Sometimes it is not easy to say whether the mutation is a result of the effect of a definite article or a preposition e.g. Waun-wen, since Welsh *gwaun* is feminine and would mutate to *waun* after the def.art. and a preposition.

The other Welsh mutations are relatively unimportant in regard to place-names but they can sometimes help in understanding particular place-name spellings correctly. The spirant mutation consists of *c, p* and *t* changing to *ch, ph* and *th* and can be caused by *a* 'and', *â* 'with, by means of' and *ei* 'her'. This explains the form *a cholva,* 'and Colfa'. The nasal mutation is *b, c, d, g, p* and *t* changing to *m, ngh, n, ng, mh* and *nh* respectively. Again, this is not common in place-name formation, but note *Tuy yn Mlaen Heyop* (see Heyope) which would be *Tŷ ym Mlaen Heiob* in standard modern Welsh, meaning 'house in the upper part of Heyope'. The importance of the mutations is shown not only in Welsh place-names but also occasionally in their effects on English place-names, note Litton.

As it stands the list is too incomplete to provide a base for a definitive statement on Radnorshire's place-names for the reasons which I set out below and in the Explanations of the Entries. The sample of a little over 200 entries is too small for confident analysis and a more extensive study, including minor

MAP 3

RADNORSHIRE English Place-Names recorded before 1400

English □
Hybrid □

names, past and present, could lead to somewhat different conclusions, for example, in regard to the distribution of the Welsh and English languages at earlier periods. The list too is an accumulation of place-names whose dates of occurrence are quite mixed. Radnorshire is not unusual in that respect but it is important to bear in mind that the analysis of such lists rarely produces any trustworthy conclusions save stating the obvious. In Map 4 I have plotted the distribution of English place-names taken from the list and it provides us with little other than an impression that English place-names are extensively found throughout Radnorshire, except the west, and that there is a slightly denser spread of English place-names in the eastern areas.

What we really need to remember is that some place-names such as Evenjobb and Glasbury are ancient while other place-names on the map are very recent, such as Crossgates and Builth Road. A more illuminating method is to plot the English place-names according to their occurrence at different periods (making due allowance for the paucity of early spellings). In Map 2 I show those English place-names which occur before 1100. Of the fourteen place-names all but one (Glasbury, probably a hybrid) are recorded in Domesday Book 1086. The distribution is not perhaps all that surprising in that it shows the place-names concentrated in that part of Radnorshire where we would expect to find them: adjoining Shropshire and Herefordshire. At that period this area was undoubtedly an integral part of those counties. Of more interest is the fact that seven of these place-names are located *west* of Offa's Dyke, the boundary between Wales and England established by Offa, king of Mercia, in the 8th century. Evidently, Offa's Dyke in 1086 had ceased to mark the border between Wales and England, a conclusion largely borne out by studying place-names in other parts of the border areas through which the dyke runs.

No doubt there were other English place-names in Radnorshire before 1100 which simply do not occur in the

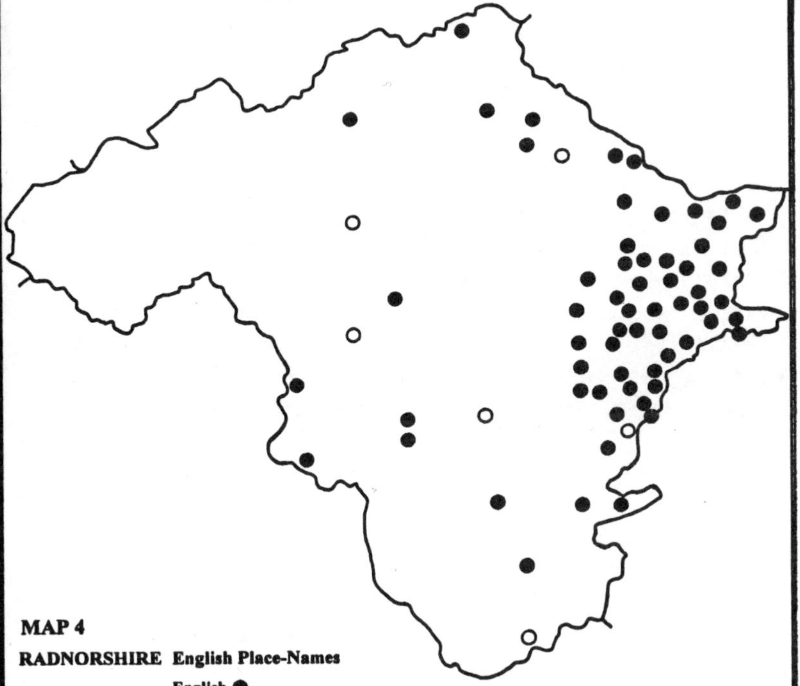

MAP 4

RADNORSHIRE English Place-Names

English ●
Hybrid and anglicised ○

surviving documentary evidence. Some of these are likely to be among the place-names shown in Map 3: those place-names recorded before 1400. Place-names such as Gladestry (first recorded c.1250) and Heyope (first recorded 1389) have an 'early appearance' and, as parishes, are likely to be older than the available dating evidence suggests. Whatever the case the distribution of English place-names in Map 3 does not show any radical difference from Map 2: the English place-names are concentrated in the same part of Radnorshire, with only a modest extension westwards (mainly in the New Radnor area). In a couple of cases we can say that some of these place-names were coined at dates subsequent to 1100: note Painscastle and New Radnor.

I have not made any attempt to plot Welsh place-names in Radnorshire partly because of the inadequacy of the evidence. Our sources are very 'lop-sided' in that they tend to come from English sources, especially the records of English government before about 1550. Welsh manuscript evidence is scarce before the late 13th century. There is, in other words, a distinct bias against the representation of Welsh place name forms in our evidence. Whether places such as Burlingjobb, Casgob or Disgoed possessed distinctive Welsh names c.1086 is simply unknown. It is likely that any such names have been lost. For this reason, we have to be careful in our deductions and it would be quite wrong to assume that eastern Radnorshire was thoroughly anglicised (on the basis of that evidence) at that time; Welshmen are actually recorded here in Domesday Book. What Welsh place-names do occur here in later periods tend to be 'cymricisations' of English place-names, i.e. English place-names with Welsh spelling and stress, as in the case of *Acyll* (Ackhill), *Einsiob* (Evenjobb), Casgob, *Gwestyn* (Weston), *Nortyn* (Norton), ? *Cinert* (Kinnerton) and *Pyllalai* (Pilleth) rather than distinctly Welsh formations such as *Trefyclo* (Knighton) and *Llanandras* (Presteigne). At least one place-name in this category seems to be a Welsh translation of an English name, e.g. *Tre'rdelyn* for Harpton.

On these grounds, I think that we can cautiously accept that eastern Radnorshire was largely English in speech c.1086 and that Welsh influence grew, both in a geographical and a social sense, from about the 14th century, reaching a peak about the 16th century or a little later. Such a pattern is replicated in adjoining areas of western Shropshire and eastern Montgomeryshire and other parts of the Marches over the same period. The Welsh expansion was a very gradual process but became sufficently dense to lead to the loss of some English place-names e.g. *Alretone* 1086 eventually became Welsh *Trewern* MTG with a similar meaning. There are like concentrations of cymricised English place-names in those areas, such as *Ffordun* (earlier *Furtune, Forton*), *Yr Ystôg* (Churchstoke) MTG, *Gwrthymp* (Worthen) and *Hobendrid (Edretehop* 1086) SHR with a few distinct Welsh place-names, e.g. *Crugion* (alias *Criggion)* MTG and *Llanffynhonwen* (Chirbury) SHR and a few of the Harpton type, e.g. *Brynedrid* (Edderton) SHR and *Trefesgob* (Bishop's Castle) SHR.

The Welsh 'advance' seems to have been reversed by the gradual process of anglicisation beginning, it would appear, in the 17th century in Radnorshire (see above under **The Welsh Language in Radnorshire**). Perhaps the present position is a salutary reminder that the distribution and pattern of place-names may not be an accurate guide to the actual language spoken in a particular area. Certainly, the impression that one gains is that the place-names and minor names are overwhelmingly Welsh yet Radnorshire is one of the most anglicised parts of Wales in regard to the spoken language. Of course, behind the apparently solid facade of good Welsh place-names such as Llanddewi, Llanbadarn Fawr and Pen-y-bont one has the anglicised pronunciations 'Lanjŵi', 'Lanbadân' and 'Pen-i-bont' (in Welsh orthography) – pronunciations which do not represent Welsh dialect – but the erosion of Welsh place-names has on the whole been relatively minor. Place-names do not change easily even when the majority language becomes a minority language.

Explanation of the entries

Firstly, I have assumed that the reader is not a specialist in place-name study. Secondly, I have assumed that the reader is not necessarily familiar with the history and geography of Radnorshire. Specialists in place-name studies will recognise the ways in which this list differs from the academic's method as seen in the publications of the EPNS, notably in the treatment of dates, the absence of source references and the limited discussion of the individual place-names. Essentially, this list is for the general reader and the local historian; if it is of interest to the place-name scholar and student then so much the better.

The area of study is the old county of Radnorshire within the boundaries of the county in the period 1889 to 1974 when Radnorshire was reduced (administratively speaking) to the status of a district authority within the new county of Powys (1974-96). The former county differed in a few minor respects from the county area which existed between 1542 and 1889: (a) the township of Litton-with-Cascob was an 'island' of Herefordshire until absorbed into Radnorshire in 1889; (b) a small area on the right bank of the river Wye (part of Glasbury parish) was lost to Breconshire in 1889; (c) an area of 'disputed ground' adjoining Cardiganshire was added to Radnorshire before 1889.

Each entry follows a set format:

(a) The name in bold letters according to its authentic spelling not necessarily that of the OS. Generally speaking, if a place has an English name, such as Newchurch, this takes precedent over the Welsh name in the alphabetical order of entries. The Welsh form in such cases is shown immediately after the 'lead form'. When a place has only a Welsh name I have preferred to employ the standard Welsh spelling, e.g. Rhaeadr (not *Rhayader*) and Glasgwm (not *Glascwm*). In a few cases I have preferred the Welsh spelling even if the place-name is of English origin, e.g. Casgob rather than *Cascob,* in those cases which have been cymricised to the extent of extinguishing the expected English form. Casgob might otherwise have

developed into an English *Cascope* or some similar spelling.
(b) The four-figure OS grid reference, except in the case of those names applying to rivers and wide territorial units.
(c) A selection of place-name spellings with dates of occurrence. Conciseness has demanded only a small sample of past spellings in the case of most entries. The dates given refer to the dates of occurrence as given in the sources not the dates of the actual manuscripts and printed sources. For example, if there is a mention of a particular place-name form in a chronicle under the year 1252 then that will be the date shown in the list not the date of the manuscript (which may have been compiled or copied at a much later date). I have done this with some reluctance since, as is well known, copyists sometimes did not preserve the original spellings of their sources. In a few cases, I have cited the manuscript date in parentheses when the date of the place-name spelling is exceptionally early (note Caerfaelog).
(d) A short discussion of each place-name. I have attempted to keep this as brief as possible and to avoid scholarly terminology such as *epenthesis, provection* and *metathesis*. When I have used these terms I have added an explanation. Many readers will be aware that the meaning of many place-names is often obscure and that explaining their development is sometimes difficult and for these reasons will readily understand the somewhat tedious caution and the ubiquitous 'possibly' and 'probably'. I have not given lengthy references to academic and other sources.

The choice of which entries to include and which to exclude was difficult but I have opted to include all those names shown on the OS 1:250,000 map covering Radnorshire published in 1989, to which I have added the names of ancient *cantrefi*, hundreds, commotes, manors and reputed manors, ecclesiastical parishes and chapelries, townships, many hamlets and the main rivers. In some cases I have combined the discussion of separate place-names under a particular entry and cross-referred, e.g. Rhydieithon see Llanfihangel Rhydieithon and Swydd Ddinieithon see Cefnllys, in order to avoid repetition.

Abbreviations

ADG	G.O. Pierce, T. Roberts a H.W. Owen, *Ar Draws Gwlad* (Gwasg Carreg Gwalch, Llanrwst 1997)
ANG	Anglesey
BRE	Breconshire
CHE	Cheshire
Coplestone-Crow:	See Bibliography
CRD	Cardiganshire
CRM	Carmarthenshire
CrF	Ffransis G. Payne, *Crwydro Sir Faesyfed* (Llandybïe 1966, 1968)
def.art.	definite article
DEPN	Eilert Ekwall, *The Concise Dictionary of English Place-Names* (Oxford 4th ed., 1960)
DNB	Denbighshire
DRB	Derbyshire
E	English
EANC	R.J. Thomas, *Enwau Afonydd a Nentydd Cymru* (Cardiff 1938)
el,-s	element,-s
EPNE	A.H. Smith, *English Place-Name Elements* (EPNS, Cambridge 1956)
EPNS	The English Place-Name Society
fem.	feminine
Field	see Bibliography
GLA	Glamorganshire
GLO	Gloucestershire
Haslam:	see Bibliography
HER	Herefordshire
masc.	masculine
MDX	Middlesex
ME	Middle English (c.1150-c.1500)
MER	Merionethshire
ModE	Modern English (c.1500 to present)

ModW	Modern Welsh
MON	Monmouthshire
MTG	Montgomeryshire
NCPN	B.G. Charles, *Non-Celtic Place-Names in Wales* (Cardiff 1938)
NFK	Norfolk
OE	Old English (c.400-c.1150)
OI	Old Irish
OS	Ordnance Survey
OW	Old Welsh (c.900-c.1200)
Parochialia:	see Bibliography
PEM	Pembrokeshire
pers.n(s).	personal name(s)
pl.	plural
pn(s.).	place-name(s)
PNANG	Gwilym T. Jones and Tomos Roberts *The Place-Names of Anglesey* (Bangor 1996)
PNDP	G.O. Pierce, *The Place-Names of Dinas Powys Hundred* (Cardiff 1968)
PNEF	Hywel Wyn Owen, *The Place-Names of East Flintshire* (Cardiff 1994)
PNPEM	B.G. Charles, *The Place-Names of Pembrokeshire* (Aberystwyth 1992)
r,-s	river,-s
RAD	Radnorshire
SFK	Suffolk
SHR	Shropshire
sp(s).	spelling(s)
SUR	Surrey
W	Welsh
WLIR	*The Welsh Language before the Industrial Revolution*, ed. Geraint H. Jenkins (Cardiff 1997)
YN	Yorkshire, North Riding

Bibliography

The following list comprises those books and articles which I have found most useful in compiling this volume on Radnorshire place-names and which I think are likely to prove of value to anyone interested in this subject and the wider subject of Welsh place-names.

Welsh History

Place-names have an historical background. Fortunately, Wales now has a new history appearing as volumes in a series. The following are most useful:

R.R. Davies, *Conquest, Co-existence and Change*. Wales 1063-1415 (Oxford and Cardiff 1987) available in paperback as *The Age of Conquest. Wales 1062-1415* (1992)

Glanmor Williams, *Recovery, Re-orientation and Reformation. Wales c.1415-1642* (Oxford and Cardiff 1987)

Note also:

A Bibliography of the History of Wales (2nd ed., Cardiff 1962) and *Supplements* in the *Bulletin of the Board of Celtic Studies* vols. 20, 23, 25.

Bibliotheca Celtica. A register of publications relating to Wales and the Celtic peoples and languages (Aberystwyth 1909-38, 1954-)

J.E. Lloyd, *A History of Wales from the Earliest Times to the Edwardian Conquest* (3rd ed., 1939)

Llyfryddiaeth Llenyddiaeth Cymraeg, ed. T. Parry and M. Morgan (Cardiff 1976) and Supplement in the *Bulletin of the Board of Celtic Studies.*

Welsh Place-names and Language

B.G. Charles, NCPN
B.G. Charles, PNPEM
Dewi Davies, *Welsh Place-Names and their meanings* (Aberystwyth no date)
Rev. Ellis Davies, *Flintshire Place-Names* (Cardiff 1959)
Elwyn Davies (ed.), *A Gazetteer of Welsh Place-Names* (Cardiff 1957)
Y Geiriadur Mawr, ed. H. Meurig Evans and W.O. Thomas, with Stephen J. Williams (1st ed. Llandybïe and Aberystwyth 1958)
Geiriadur Prifysgol Cymru (Cardiff 1950-)
Geraint H. Jenkins (ed.), *The Welsh Language before the Industrial Revolution* (Cardiff 1997)
Bedwyr Lewis Jones, *Enwau* (Capel Garmon 1991)
Gwilym T. Jones and Tomos Roberts, PNANG
Anthony Lias, *Place-Names of the Welsh Borderlands* (Ludlow 1991)
J. Lloyd-Jones, *Enwau Lleoedd Sir Gaernarfon* (Cardiff 1928)
Allen Mawer, *A Survey of the Place-Names of Wales* (1932-3)
John Morris-Jones, *A Welsh Grammar* (Oxford 1913)
Rev. Thomas Morgan, *Handbook of the Origin of the Place-Names in Wales and Monmouthshire* (Merthyr Tydfil 1887). [This needs to be followed with great caution]
T.J. Morgan and Prys Morgan, *Welsh Surnames* (Cardiff 1985)
Melville Richards, *Welsh Administrative and Territorial Units* (Cardiff 1973)
Hywel Wyn Owen, PNEF
G.O. Pierce, T. Roberts a H.W. Owen, ADG
R.J. Thomas, EANC
Ifor Williams, *Enwau Lleoedd* (Liverpool 1969)

British and English Place-Names

Kenneth Cameron, *English Place-Names* (London 1961; 3rd ed. 1977)
Bruce Coplestone-Crow, *Herefordshire Place-Names* (BAR British Series 214, 1989)
John Field, *English Field-Names. A Dictionary* (Gloucester 1989)
Eilert Ekwall, DEPN
Eilert Ekwall, *English River-Names* (Oxford 1928 and 1968)
Margaret Gelling, *Place-Names in the Landscape* (London 1984)
Margaret Gelling, *Signposts to the Past* (London 1988)
Margaret Gelling, W.F.H. Nicolaisen and Melville Richards, *The Names of Towns and Cities in Britain* (London 1970)
P.H. Reaney, *The Origins of English Place-Names* (London 1960)
Adrian Room, *A Dictionary of Place-Names in the British Isles* (London 1988)
Adrian Room, *A Concise Dictionary of Modern Place-Names in Britain and Ireland* (Oxford 1983)
A.H. Smith, *English Place-Name Elements* (EPNS, Cambridge 1956)

Radnorshire

Richard Haslam, *The Buildings of Wales. Powys* (Harmondsworth and Cardiff 1979)
William Henry Howse, *Radnorshire* (Hereford 1949)
Llywelyn Hooson-Owen, 'The History of the Welsh Language in Radnorshire since 1536', MA thesis (School of Celtic Studies, University of Liverpool n.d. [1960s])
Fransis G. Payne, CrF
Lord Rennell of Rodd, *Valley on the March. A History of a Group of Manors on the Herefordshire March of Wales* (Oxford 1958)
Rev. Jonathan Williams, *A General History of the County of Radnorshire* (c.1818-21), ed. Edwin Davies, Brecon 1905)

Note also the articles which have appeared in the *Transactions of the Radnorshire Society* (1931-)

Abbeycwmhir : Abatycwmhir SO 0593
Cumhyr c.1200, Cumhir 1254, y Kwmhir 1198, Abby Cwmheare 1752

A modern n., coined in the 18thC, probably about the time that Sir William Fowler erected St Mary's church (re-built in 1866) near his mansion Abbeycwmhir Hall and the ruins of the abbey of Cwm-hir established by the Cistercians in 1143. The valley (*cwm*) is narrow and twisting rather than 'long'. Lewis Morris 1796 called the abbey *Mynachlog y Cwm Hir* (W *mynachlog* 'monastery') but most W sources call it simply *Y Cwm Hir*, 'the long valley'.

Aberedw SO 0847
Aberedo c.1291, Aberedo c.1385, Aberedou 1293, Abreduwy 1293, Aberedwy 1375

The first el. is the common W *aber* 'mouth of a river, confluence' since the parish lies at the mouth of the r. *Edw* or *Edwy (Edw, Edwy* late 15thC) but it is difficult to say what this means unless it has W *edn* 'bird, fowl', and 'fly, bee' as R.J. Thomas (1938) suggested for the r. *Eden* MER. *Edw* might mean something like 'flying river', i.e. a river which dips and twists like the flight of a bird.

The traveller and antiquarian John Leland mentions *Aberhedon Castel* here c.1536-9 though this has been misidentified with Brecon (W *Aberhonddu*). The castle is a small mound near the r. Wye, partly destroyed by the construction of the railway from Builth Road to Three Cocks Junction in the 19thC. Haslam records that it was built of stone and is likely to date from c.1284.

Ackhill : Acyll SO 2865
Achel 1086, Achul 1293, Akhull 1304, Akelthe 1546

OE *ǣc* and *hyll* 'oak-tree hill'. The actual hill may be Furrow Hill. The W name is implied by the 1546 spelling since E speakers sometimes substitute *thl* for *ll* in speech; many readers will be familiar with the E mispronunciation *Dolgethli* for *Dolgellau* MER. W speakers were just as liable to substitute a familiar sound, in this case putting *-yll(t)* in place of E *-hill*. *Gwersyllt* DNB and *Sodyllt* SHR are thought to show the same development.

Ackwood SO 2365

Acwode 1304, Ackewode 1340-1, Acwoddslond 1383-4, Ackwood 1580

OE *ǣc* and *wudu* meaning 'oak wood'. The last but one reference means 'land belonging to Ackwood'. I have not seen a W n. for Ackwood.

Bache Hill hill SO 2163

Black Hill 1815, Bach Hill 1832

The hill is near a farm called Bache (SO 227628) spelt *Bach* 1862. This is likely to be OE *bæce, bece* 'a stream, a valley' found widely in neighbouring parts of England as *-bach, -batch, -beach*. The 1815 spelling is probably an error.

Badland SO 2363

Badelonde 1304, Badlond 1487, Badland 1587, (wood) *Badlondesshelde 1304*

OE or ME meaning 'Badda's or Bada's land' or perhaps simply 'bad land, land of little worth'. One sp. has *helde* from OE for 'slope'.

Bailey Hill hill SO 2372

BAILY HILL 1815, BAILEY HILL 1833

The n. appears to be E though *Bailey* may be W *beili* meaning 'bailey, yard', occurring commonly in mid Wales in the ns. of farms. It might also have the sense of 'farm'; the word, of course, is borrowed from E.

Barland SO 2862

Bernoldune 1086, Beraldon 1304, Bernaldon 1343, Barlenton 1452, Barlondon 1556-8, Barlond 1544

'Beornweald's hill', with OE *dūn*, a very common pn. element in England for 'a hill', in this case referring to the unnamed hill S. of Barland farm. The pn. has changed in two ways: firstly by metathesis, i.e. *-nal-* or *-nol-* has become *-len-* or *-lon-*, and secondly the pn. has lost its 'tail'. The second change may be the result of W influence since W words of more than two syllables are usually stressed on the penultimate, e.g. *Carnéddau.* When an E pn. is borrowed by W speakers two things often happened: either they shifted the stress to the 'correct W syllable (try stressing *Huntingdon* or *Birmingham* on the *-ing-*) or the stress remained on the first syllable as it usually is in E, and the unstressed third syllable was dropped. This second change seems to have happened here (*Huntingdon* and *Birmingham* would become *Hunting* and *Birming*) and there are similar examples in other parts of Wales and SHR where W was once spoken.

Beacon Hill hill SO 1776

Beacon 1836

A beacon is marked here on *BEGUILDY MOUNTAIN* on the OS 2" scale map drawings 1820-1 but it is not necessarily proof of a beacon since the word was sometimes applied to a prominent hill. It might also refer to four barrows here (Haslam, p.220).

Beguildy, see **Bugeildy**

The Begwns hill SO 1544

Beacon Hill 1817, BEACON HILLS 1828, The Begwns 1833

Compare Beacon Hill. OE (*(ge)bēcon* 'sign, signal, beacon' was borrowed in W as *begwn, began* and occurs in *Twynbwmbegan* (Wenvoe) GLA and perhaps the hill-name *Pegwn-mawr* and

Pegwn-bach (Llandinam) MTG but it is a little difficult to distinguish it from *pegwn* 'pole' and even *pigyn* 'thorn, tip' (note *Penypigyns* and the farm *Pikins* SN 9498 near Carno MTG). The sps. may well represent the W and E ns. for the hill.

Beggarsbush SO 2664

Beggars-bush 1698, Beggarsbush 1815, Beggars Bush 1828

Similar ns. in England applied to fields appear to mean 'land of poor quality' *(Beggar's Furlong, Beggar Ground* etc; see Field (1989); a mocking n.

Betws Cleiro SO 2247

y bettws c.1566, Bettus 1670, the Bettws in Clerrowe 1615

W *betws* is a borrowing from OE *bed-hūs*, ''house-of prayer' and often had the sense of 'chapel-of-ease' as in this case; it lay in the parish of Cleiro.

Betws Diserth SO 1156

Betws yn elfel c.1562, y bettws c. 1566, The Bettus 1546, Bettus Dissard c. 1700

See Betws Cleiro – in this case a former chapel-of-ease in Diserth parish. Edward Lhuyd, the antiquarian and scholar. c.1700 refers to *Bettws bach* ('little Betws') which may be this place. Betws Diserth lay within the ancient *cantref* (or hundred) of Elfael, hence the first sp.

Bilmore SO 2558

(John) Billemor 1338, (messuage) Bilmor 1591, Byllmor 1563, Bilmore otherwise Bilmore Farm 1760

OE meaning 'Bila's moor'; the n. of a manor otherwise called Stanner.

Black Mountain hill SO 1677

Black Mountain 1820-1, 1833

A 'dark mountain' perhaps because of its rocks or shadows.

Bleddfa SO 2068

Bledwach 1195-6, Bledvach 1262, Blethevagh 1304, Blethevach 1409, Blethvah 1695, Bleddfach 1600-7

Probably W *blaidd* 'wolf' and *bach* 'nook, angle'; compare Wolfpits. The first el. could also be a pers.n. *Blaidd*. The end of *Bleddfach* has been lost, probably under E influence, as in the case of *Solfa(ch)* PEM.

Bochrwyd, Boughrood SO 1239

Bouret 1205, Boskeret 1206, Bocred 1242-3, Boghred 1234, Bochroed c.1291, Boghroyde 1513, bochrwyd c.1566, Bochrwyth 1400

A difficult pn. Later spellings from W sources suggest that it may be W *boch* 'cheek, jaw' and *rhwyd* 'net, snare' and perhaps it is not too fanciful to favour 'jaw-shaped net or fish-trap (in the r. Wye)', but I cannot think of any similar pns. Some later W spellings have *-rwd* showing the spoken form, just as *diffwys* 'steep slope' etc. is often pronounced 'diffws'.

The early sps. are odd and an E explanation would be possible, i.e. 'beech clearing', with OE *bōc* and *rōd* (as in *Boughton* and *Huntroyde* in DEPN) were it not for the numerous W forms (*Bochrwyd, -rwid* 1574, *Bochroyd* 1549, *Bochrowd* 1620 etc.). Modern maps spell the pn. *Boughrood* but locals pronounce it 'Bockrood' or 'Bochrood' (with *Boch-* as in Scottish *loch*).

Bronllys SO 1071

Brwynlhys 16thC, Bruntlis 1688, Brointlis 1670

Probably W *brwyn* and *llys* meaning 'hall constructed of rushes', not to be confused with *Bronllys* BRE which is *Brendlais, Broynles* late 12thC, *Breinleis* 1210-2, *Brentles* c.1291, *Broynles* late 12thC, *Bryntheles* 1322 and *Brwynllys* 16thC which appears to mean the same or 'court of a man called Braint'. In both cases the first els. have become confused with W *bron* (as in Bronydd).

Bronydd SO 2245

Bronreyht 1195-6, y Bronnydd 1682, Bronydd 1833

The identification of *Bronreyht* with this place is not certain and is at odds with later sps. which favour *bronnydd* the pl. of W *bron* in its common topographical sense of 'hill' (it also means 'breast' and 'near to'). Bronydd stands at the foot of a hill overlooking the r. Wye.

Bryndreiniog, Bryndraenog SO 2078

Brynn dreinyoc c.1600, Brendrenog 1559, Bryn Draenog 1820-1, Bryndraenog 1833

'Thorny hill', with W *bryn* and *dreiniog, dreinog* not *draenog* which usually means 'a hedgehog'. The root of both words is found in *draen, -en,* pl. *drain,* 'thorn, thorn-tree'. Bryndreinog is best known for its ancient timber hall-house (see Haslam, 219-20.

Bryn-gwyn SO 1849

Brewen 1231, Bringwyn c.1291, Bryngwyn 1361, ll.V'el y bryngwyn c.1566

'White or fair hill', with W *bryn* and *gwyn,* presumably referring to the paleness of the soil or rocks. The church is dedicated to St Michael (W. *Mihangel*), hence the W form c.1566 standing for *Llanfihangel y Bryn-gwyn*. Local people pronounce the pn. *Bryng-wyn,* roughly 'Brung-win' in E spelling.

Brynhyfaidd SO 1066

Brynhyfedd 1791, Bryn-rhyfedd 1833, Bryn Hafod 1820-1, Brinbeneth 1670

'Hill associated with a man called Hyfaidd'. The pers.n. became *-hyfedd* in local dialect but its unfamiliarity to some writers made it vulnerable to misspelling; compare *Maesyfed* (see New Radnor). The 1820-1 sp. has confused the n. with W *hafod* 'summer dwelling'.

Bryn Llyndwr hill SO 0683
Llyn dwr Hill 1817, 1836

Named from *Llyn-dwr* 1836 which appears to be W *llyn* 'lake, pool' and *dŵr* 'water' but it is a suspicious combination and earlier spellings are needed.

Brynseifiog SO 1441
Bryn siviog 1685, Bryn Siviog 1654, Brynsivioch Manor 1647, Brinseviock, alias Solervaur 1740

Apparently, 'hill abounding with strawberries', with W *bryn* 'hill' and *syfïog* (*syfi* 'strawberries') but the hill seems too high and exposed for cultivating such fruit and the n. may be fanciful (like *Strawberry Hill* MDX; see Field, 221) rather than descriptive. There is a *Cwmsyfïog* near New Tredegar MON.

Bryn Titli hill SN 9375
No early sps. seen but it may be named after nearby *TITTLEY MILL* 1835 which may contain the E surname *Titley*, itself derived from the pn. in HER (for which see DEPN and Coplestone-Crow, 190).

Bryn y Maen hill SO 1657
Bryn Maen c.1817, Bryn-y-maen 1833

'Hill distinguished by a stone or boulder', with W *bryn*, def.art. and *maen* 'stone'.

Bugeildy, Beguildy SO 1979
Bugeylde c.1291, Bugeilde 1397, Bygeldy 1392-3, Bugelte in Malenhith 1259, ll.V'el y bigil dy c.1566

'Shepherd's house or hut', with W *bugail* 'shepherd'. The valley west of the village is *Cwmbugail* (so called 1833), i.e. 'shepherd's valley', though it may be a contracted form of *Cwm y Bugeildy*. The church's dedication to St Michael is recorded in 1259 and this explains the c.1566 spelling (*Llanfihangel y Bugeildy* in ModW spelling), probably a 'bookish' form since most sources give the short form *(y) Bugeildy*. The def.art. has been lost; it occurs in corrupt form in *Abergulde* 1528 confused with W *aber* 'mouth of a river, confluence'.

Builth Road see **Cwmbach Llechryd**

Burfa SO 2761

Borewere 1383-4, Boreware 1395, Barvar, -vor *1649, Burward 1667, 1670, Barwarr otherwise Burvarr Farm 1760*

Possibly OE *bor* and *wer, wær* meaning 'upper weir' or 'weir by a hill'; the 'hill' may be *Burfa Camp*. The alternation of *w* and *f* (= E *v*) is found in W (see Doldowlod).

Burlingjobb SO 2558

Berchelincope 1086, Bertlinghope 1304, Byrchop 1538-44, Berchoppe, Beroschoppe, Berlingshopp or Berlingobb 1634

'Valley associated with a man called Berhtel'; an OE pers.n. with the compound *-inghop* often found in E pns. OE *hop* – usually found as *hope* – had a number of meanings including 'enclosed valley' and also 'settlement in remote valley' which seems to be its commonest sense in Radnorshire and adjoining parts of England. The original n. was contracted to *Byrchop, Berchoppe* but this pronunciation no longer survives. The manor here is called *Berling or the Diche* (? *on the Diche*) 1543': the 'ditch' refers to Offa's Dyke about two miles to the east.

Bwlchysarnau SO 0274

Bulch y Sarny 1755, Bwlch Sarne 1781, Bwlch y Sarny 1813, Bwlch-y-sarnau 1820-1

W *bwlch* 'pass, gap', def.art. and *sarnau* 'causeways' referring to the roads here crossing the boggy uplands around the hamlet.

The OS map 1833 misinterpreted the name as *Bwlch y carnau*, with W *carnau* 'cairns'.

Cabalfa SO 2346

(grange) *Cubalva 1231, Kenbalva 1232, Kybalva 1397, Cabalva in Elvel 1535, Inis y Gabalva 1562-3*

W *ceubalfa* 'ferry' as in the case of *Gabalfa* (Cardiff). The ferry across the r. Wye connected the Cleiro-Whitney road with Clifford HER. Cabalfa was in the *cantref* or hundred of Elfael. The 1562-3 reference has W *ynys* which means 'river-meadow' as well as 'island' and shows the mutation of the fem. word *ceubalfa* after the def.art. (see **Introduction**).

Caban Coch Reservoir see **Elan**

Caerfaelog SO 1073

(battle) *gart mailauc, – Garthmailauc c.722 (c.1000, c.1286), gard uaelawc, gard uaeloc 15th C., Carviloge 1771, Caerfaelog-isaf, – uchaf 1833, Ardd Faelog 1633*

Probably 'enclosure associated with a man called Maelog', with W *garth* or *gardd* applying perhaps to the enclosure where ruins of a 'chapel', and a baptismal font, were recorded in 1805. It is possible that the ruins are really those of the mansion (*plas*) of Maredudd Fychan (15thC) visited by many W bards such as Llawdden, Ieuan Swrdwal and Lewys Glyn Cothi (see CrF II, 75). It is impossible to say whether or not the *Maelog* in this pn. has any connection with the saint *Tyfaelog* recalled in *Llandyfaelog Fach* and *Llandyfaelog Tre'r-graig* BRE (see Llandrindod). The pn. survives in corrupt form because *Gar(dd)-* has become confused with *Gaer-*, i.e. W *caer* 'fort'.

Cantal SO 2249

Cantal late 16thC, Cantal, Kantalhall 1820-1, Kantall 1644, Cantel 1771

Unexplained; there is an obscure Brit. word **canto-, *canti-* found in pns. and tribal ns. of the Roman period which has been linked with W *cant* 'edge, border, enclosure' and 'host, party' (EPNE I, 90) but it is very difficult to say what Cantal actually means if it has this element. W *cantel* 'rim, brim' seems to be a late borrowing from E *cantle* and fails to match most of the sps. *Cantal* might be W *cant* 'a hundred (the number)' and *tâl* 'front end' (compare *cantref* from *cant* and *tref,* lit. 'a hundred townships') but I do not know of any similar pns.

Carnaf SO 2249

Caernaf 1231, Karnaff 1232, Carnaf 1735, Karnayth in Elvell 1397

The n. probably survives in *Carney* Cottage (*Caernay* 1816-17) and formerly referred to a grange or farm of Cwm-hir abbey, identified by Haslam and others with Court Farm though this is uncertain. The pn. may contain W *carn* 'cairn, barrow; rock pile' and a suffix *-af (-aw, -o)* found also in *Carno* MTG. Both Carno and Carnaf have Roman forts, in the case of Carnaf near the house *Gaer* near Cleiro, but the sps. of *Carno* generally have *-o, -aw* not *-af, -aff.*

The 1397 sp. with *-th (= -dd)* may show the same sort of dialect confusion of *-f* and *-dd* that occurs in *Caerdydd* (= Cardiff), earlier *Caerdyf.* For *Elvell* see Elfael. The grange, or at least part of it, may have been at *Great Tirymynech* (so spelt 1816-17), now called *Boatside,* a n. seemingly transferred from the farm at SO 222430 *(Boatside Fm 1816-17),* or *Little Tirymynech,* both meaning 'the monks' land' with W *tir* 'land', def.art and *myneich* an old pl. of *mynach* 'monk'.

Carneddau hills SO 0655

Carneddau c.1817, 1833

Pl. of W *carnedd* 'heap, tumulus'; a farm on their eastern side has the same n. and it is unclear which bears the original n. but the hills are very lumpy in shape like heaps of stones.

Carogau SO 1075-6

o Garog 1597, Krough 1581, Caroge 1678, Carogey 1670, Croga 1670

W *carogau* pl. of *carrog* 'torrent, stream' though local people call the two farms 'Carôgi – compare some of the sps. of Bwlchysarnau and Dolau. The W pl. ending *-au* is variously pronounced *-e* and *-a* in other parts of Wales but *-i* is likely to be a product of anglicisation. The farms are called *Crogau-fawr* and *-fach* 1833, i.e. 'Big' and 'Little Carogau'. The first two sps. have inadvertently omitted a final *-e* or *-a*.

Casgob Cascob, SO 2366

Cascop, -e 1086, Cascop 1304, Cascob 1343, Cascob, -cobb 1670

OE, probably meaning 'valley associated with a person called Casca', with *hop* as in Burlingjobb. It is unlikely to have E *cask* 'barrel' which appears to be a late addition to the E language although it is curious that the adjoining dingle is *Cwm y Gerwyn* (*Comm'geron* 1325-7, *Cumbegerewyn* 1360-2). This seems to mean 'valley of the vat' or 'valley shaped like a vat', with W *cwm,* def.art. and *cerwyn*.

Castell Collen SO 0562

Castell Colwyn c.1607, - - Collwyn 1709, Castell-collen 1833

This may be *Castell gollwin* 13thC. The first el. is W *castell* 'castle' and the second appears to be *collwyn* 'hazel-bush, hazel-wood' rather than *collen* 'hazel-tree'. The antiquarian Fenton 1804-13 calls it *Caer Collen* (otherwise unattested in early spellings) and says it refers to the Roman auxiliary fort here; he says that a local landowner, Williams of the Cwm, derived the n. 'from the number of Hazels growing over the sides of it'.

Castell Crugeryr SO 1559

Castrum Crukeri 1188, or grygeryr c.1460, Tommen Krygerydh c.1700, Bryn-Y Penn alias Cregereth 1713, KREGERIR, -erer 1602, Crykereth Grange 1544

The n. of a small medieval castle: W *crug* 'mound', often 'castle mound', and *eryr* 'eagle'. Some spellings have the variant *-erydd* for *-eryr*. *Castell* seems to be a fairly modern addition to the pn.; the c.1700 form has W *tomen* 'heap' which often occurs in W castle ns.

Castell Dinbawd SO 0975

Dinbot 1281-2, Dymbaud 1304, Dyneboth, Tynbot 1536-9, C(astell) Tinbod c.1600, Castell Tinbott, Tibbot or Timboth 1813

A difficult pn.; it appears to be W *din* 'fort' and *bod* 'house, dwelling' but a better suggestion (of Prof. G.O. Pierce) is that it could be a variation of W *twmpath* 'hillock, mound', a word often applied to castle mounds (like *crug* and *tomen*; see Castell Crugeryr). Several minor ns. in GLA spell the word *dimpath, dinbath*. It is possible that some sps. of *Dinbawd* have been influenced by knowledge of the common elements *din* and *poeth, boeth* 'burnt'. I have not seen any explanation of *twmpath*; it may be a borrowing from ME *tump* 'a hillock, mound' etc. found mainly in the west Midlands and western England but this is said to be of obscure origin (*Shorter Oxford English Dictionary*).
There are fanciful tales deriving the pn. from the pers.n. of *Maud* de St. Valéry, wife of William de Breos (in the time of King John) or *Maud*, wife of Roger Mortimer (died 1282), lord of part of RAD.

Cefnllys SO 8861

Keventhles 1246, Kenlles 1322, Kevenlles 1281-2, Kefn y llys 1272-4, ll.V'el kefen llys c.1566

W *cefn* 'back, ridge' and *llys* 'mansion, court' – the site of a medieval castle, church and failed borough. The castle stands at

one end of an abrupt, isolated hill which also bears the ruins of an Iron Age fort, thought to have been called *Dinieithon*, i.e. *din* 'fort' and the n. of the r. Ieithon. This occurs in *bro dineithon* c.1170-9, (castle) *Dynyeythan* late 11thC, *Soyth in Eythan* 1332, *Comm. Soedinethon* 1536-9, with *bro* 'country, lowland' and *swydd* 'subdivision of a *cantref* or hundred, commote'. Swydd Dinieithon was the area centred on Cefnllys. The pn. was first contracted to *Swyddineithon*, as shown by the 1332 and 1536-9 spellings and finally the unstressed *-i-* was lost.
Between the castle and the river Ieithon (Ithon) is the church of St Michael; this explains the c.1566 sp. which stands for *Llanfihangel Cefnllys* 'church of St Michael at Cefnllys'; compare Llanfihangel Helygen and Bugeildy.

Cefnpawl SO 0670

Kevenpawl 1232, Keven Ipaulle 1541-2, Kevenpoole 1670, Cefn-pawl 1820-1

W *cefn* as in Cefnllys and *pawl* 'post'; the hill here, now called *Llywy* (*Llowey* 1820-1, 1833), may once have borne some sort of post or pole. There is another *Cefn-pawl* in Bugeildy which occurs as *Keven y nawl* c.1700 and *Cefn Pawl* 1820-1, evidently with the same meaning despite the c.1700 spelling (in a badly edited text). *Llywy* itself may take its n. from a stream since *llywy* means 'bright'.

Cefnsuran SO 2271

Keven Sweren c.1700, Kevensuran 1751, Cevn suran 1833

W *cefn* as in Cefnllys and *suran* as in *suran y coed* 'wood sorrel'. The c.1700 spelling is unreliable (see Cefnpawl for a similar error).

Cellws SO 0763

Kellowse 1631, Cellhous 1679, Kettlehouse 1812, Kellus 1670, Cellws 1833

Uncertain; the n. is reminiscent of words borrowed in W from E, such as *betws* (see Betws Cleiro), *tanws* 'tanhouse' and *bacws* 'bakehouse', in which case we may have W *cell* 'cell, bower' and *-ws,* but our sps. are rather late. Another possibility is W *cyll* 'hazeltrees' and *-os,* an ending often found with plant ns., meaning 'place abounding in hazel-trees'.

Cilfaesty Hill hill SO 1284

Faesdre Tump 1836, Cae Fisty 1817

The correct form may be *Cilfeisty,* i.e. W *cil* 'nook' (and often 'source of a river'), *maes* and *tŷ* 'house'; *maes* is a common pn. element meaning 'open-field' (now plain 'field') and 'outside', and taken in conjunction with *tŷ* it may mean 'house in the open-field, house in the unenclosed area', but our spellings are late and vary too much. The first form may be a result of confusion by a map-maker with *Cefnyfaestre* (SO 1290) about four miles to the north. Sps. such as *Kevn y vastre* 1739, *Keven y vastrey* 1739 and *Keven y faestre* 1630 support the belief that this means 'ridge of the open township' or something very similar; modern W *maesdref (maes* and *tref)* means the rather different 'suburb'. Cefnyfaestre is described as a township in 1603.

Cil-gu SO 0162

Kelygie 1544, Kil y gu 1648, Kylugne 1688, Kilgee 1670, 1812, Lower –, Upper Cilyci 1833

A n. misconstrued as having W *ci* 'dog' in the 1833 spelling but it seems to be W *cil* (as in Cilfaesty Hill) meaning 'a retreat, corner, nook' and *cu* 'dear, beloved'. If we are right then the pn. ought to be *Cil-cu* since *cil* is masc., but instances do occur where the following adjective has mutated as if the noun *(cil)* were fem. Some spellings of *Cwm-gu* (SO 1921) BRE are reminiscent of *Cil-gu*: *Commegy* 1561, *Cwmby Gee, Cwm y Gu alias*

Cwmgu 1792-3, and indeed *cwm* is masc. as well. The def.art. seen in some sps. of both pns. may be intrusive. Both pns., however, need further investigation; *cu* occurs in a number of pers.ns.such as *Dyddgu, Mygu* etc.

Ciltwrch SO 1540

kil twrch c.1700, Cil-twrch 1832

'Boar's nook' or 'source of stream called Twrch (the boar)'; *Twrch* may have been an old n. for the stream here, which flows down to the Wye at Glasbury, since there are similar rns. in other parts of Wales and England (note *Turkdean* GLO, in DEPN). There is also a farm *Cwm-twrch (Cwmtwrch* c.1817) near Glasgwm having the same el.

Afon Claerwen

Clayrwen 1548, Clarwen c.1755, River Clarwen 1828, Afon Claerwen 1833

Probably 'clear, or bright, white (river)', with W *claer* and *gwen*, the fem. of *gwyn* 'white'. Its main tributary is *Afon Claerddu ('Clardue,* that is to say *Blak Clare'* c *1536-9)* which has *du* 'black'. The lower portion of Claerwen's course is now part of Claerwen Reservoir constructed in 1946-52.

Clas Garmon SN 9773

Glastkarmon 1544, Glase Hermon 1670, Klasharmon 1579, Clase 1851

W *clas* meant 'monastic community, convent' and occurs quite commonly in Wales (see Glasbury); no doubt it refers to the *clas* at Saint Harmon but *Clas Garmon* or *Clas Harmon* is likely to mean not '*clas* of St Garmon' but '*clas* in Saint Harmon (parish)' – that may explain the forms with *Hermon, -harmon*. It could well be a contracted form of *Clas Sanharmon*. A similar example may occur in the adjoining parish of Llangurig MTG where we have *CLASSE KYRRICK, y classe 1560-1, Clase 1599-1600, 1608-9* meaning not '*clas* of St Curig' but '*clas* in Llangurig'.

Clatretune (lost)
Clatretune 1086

Possibly now Whitewall Farm (SO 326640) according to Lord Rennell of Rodd, meaning 'settlement or farm at Clatter Brook', with OE *tūn*. The stream occurs as *Claterbrooke* 1509, *Clatter-, Clatebroke* 1545 and may mean 'stony stream' or 'noisy stream', with OE *clater* and *brōc*, but there are similar rns. elsewhere in Wales which have *caled* 'hard, rough' and *dwfr, dŵr,* usually found as *Clet(t)wr* and the possibility has to be borne in mind that OE, ME *clater* has replaced a W n.

Cleiro, Clyro SO 2143
Ruyll 1232, Royl 1283, 1331, la Royl 1334, CLIRO sive ROYLL 1491, Clereho 1536-9, Clyrow 1549, Gleirwy 1447-86, Kylirw c.1566

Named from the r. flowing through the village, a n. composed of W *claer, clair* (as in *disglair* 'clear' and Claerwen) and the common suffix *-wy* typically found in rns., often shortened to *-w, -o*. The earlier n. *Royl* etc. has been interpreted as W *Yr Heol,* 'the road', thought to occur in *Pen-yr-heol* ('top of the road') and other pns. in this area, and there are some extraordinary theories about reputed Roman roads and their routes based on this supposition. *Royl* may actually be W *rhwyl* 'court, palace', possibly referring to the small castle here but I am unaware of any other pns. in Wales which contain this word and this suggestion must remain unproven.

Clud see **Radnor Forest**

Coedglason SO 0767
Coyid Glasson 1533-5, Coedglasson 1617, COED GLAS ONN 1811, Coed-gleision-isaf, -uchaf 1820-1

W *coed* 'trees, wood' and a pers.n. *Glason* misconstrued in two of the cited forms as 'blue, green ash-trees' 1811 and 'blue-green trees' 1820-1. *Isaf* and *uchaf* are 'lower' and 'upper' distinguishing two farms. This may be the place recorded as

Glassawn Argoedyd in the Red Book of Hergest c.1400 in company with *Tir Mael* (=Maelienydd) and *Glan Gwy* ('bank of the Wye'); W *argoedydd* may simply mean woodland.

Coed-swydd SO 1464

(wood of) *Soith 1297,* (manor) *Coidsoyth 1603-4, Coidsoith 1569-70, COYED by SOUTHE 1574*

W *coed* 'trees, wood' and *swydd* 'subdivision of a *cantref* or hundred'. The n. now applies to a small hill near the farm *Swydd* and could simply mean 'Swydd wood'; see Swydd a'r Graig.

Colfa SO 2053

I golua 1447-89, acholva (= a cholva) c.1541, Colvay 1553-5, Colva 1550, Colfa c.1817

Very uncertain and attempts to connect it with F *col* 'pass' are unconvincing since there is no evidence that the word was borrowed in W. The final el. is likely to be W *-fa* (*man*) 'place'. It may have W *colf* as in *colfen* 'branch, bough; tree' or col, *coly, cola* 'awn, beard of husks or chaff'. This may seem unlikely but there is a *Bwlchycibau* (SJ 1717) MTG *(Bwlch y Kibbe* 1764*), Bwlch-y-cibau* 1820-1, 1836*)* which may be translated 'pass of the husks' (with W *bwlch* 'pass, gap', def.art. and *cibau,* pl. of *cib, cibyn* 'shell, pod, husk'). There is also a farm *Cola* near Llangamarch BRE but I have seen no early spellings.

Colwyn

(hundred, honour) *Colewent 1265, Coluin 1215, castell Colwyn 1144,* (commote) *Colwyn 1578, Llys Colwyn 1447-89,* (castle, manor) *Colwen in Elvell 1397, Colein 1397*)

The n. of a former commote in the *cantref* or hundred of Elfael and a castle taking their n. from the r. *Colwyn,* a tributary of the Edw. The rn. may be W *colwyn* 'whelp, puppy, cub', alluding to its sounds or lively nature. There are several rs. in Wales with this n.; one, for example, gave its n. to *Colwyn Bay* and *Old Colwyn* DNB.

Craig Goch Reservoir see **Elan**

Cregrina SO 1252

kreic urna, - vuruna c.1160-1220, Crugrima c.1291, Crugryna 1513, yn Nghraig Runa 1447-86, Cregrina 1491, 1572, Keric Ryna c.1566

W *craig* 'crag, rock' and a pers.n. *Muruna*. The original form may have been *Craigfuruna* which became *Craig'runa* and finally *Cregruna* or *Crugraina, Cregraina*. Local people pronounce the n. as if it were *Cregraina* but there seems to be no evidence to show whether this is ancient.

Crossgates SO 0864

Cross-gate 1833, Cross gates 1875

Named from the toll-gate shown on the 1833 OS 1" map of the area; the village grew up around the crossing of the Builth-Newtown and New Radnor-Rhaeadr roads.

Crugybyddar SO 1583

Kruk Bechar 1304, Crukadyr 1332, (mill) *Crutpadern 1392-3, (hendy) Y Cruc Bydhar c.1600. Griygbyder 1706, Crigbother, Crigebother 1670*

Properly *Y Crugbyddar*, ie. W *crug* 'hillock, mound' referring to the castle mound *Crugyn Tump* (SO 168813) at Felindre, and *byddar* 'unable to hear, silent'. *Byddar* can also mean 'deaf person' as in *Llanybydder* CRM *(with llan)* and *Tyn Cilybyddar* (*Tin-kile-y-buddar* 1769) in Llangynllo (with *tyn*, a short form of *tyddyn* 'homestead, cottage', *cil* as in Cil-gu and def.art.). It could simply mean 'silent mound', a good n. for a deserted castle.

Cwmbach Llechryd SO 0254

Llechryd early *16thC, Llechrydd 1553-5, Court Llechridd c.1755,* (messuage) *Tir y Courte in Lleghred 1616, Cwm bach 1820-1*

Two distinct pns. which were combined for the parish created in 1887. *Llechryd* is W *llech* 'slate, slab' and *rhyd* 'ford' referring to a passage over the afon Dulas not the Wye. Near the ford is a fort, probably Dark Age, adjoining Court Farm, called *Cwrt Llechryd* 1804-13 and *Cwrt* 1820-1. W *cwrt* means much the same as E *court*. The whole area is now usually called *Builth Road*, from the railway station serving Builth BRE. *Cwm-bach* is about half-a-mile along the A479 towards Rhaeadr and takes its n. from the farm meaning 'little valley', with W *cwm* and *bach* 'little, small'.

Cwm-brith SO 0960

Cwmbreeth 1686, Cwm-brith 1833

W *cwm* 'valley, combe' as in Abbeycwmhir and Cwmbach Llechryd and *brith* 'marked with different colours (esp. black and white or red and white)' (GPC), probably indicating a valley with soil or rocks of sharply differing colours.

Cwm-geist SO 1771

Cwm-y-Gaist 1771, (meadow) *Cumb y guiest 1613, CWM Y GAST 1811*

W *cwm* as above, def.art. and *geist* pl. of *gast* 'bitch, female dog'. There is a tendency, very noticeable in RAD and BRE, to drop the def.art. in pns. of three or more elements. There is a *Cwm-gast* MTG.

Cwm-gila SO 2671

Cumb gilla 1613, Cwmbgill 1616, Comegilla 1575, Combe Gila 1704, Comeguiltha 1670

W *cwm* and perhaps an E pers.n. *Gilla,* originally Irish. Some spellings, especially that for 1670, suggest that the correct W sp. should have *-gilla* with the usual W pronunciation of *ll*.

Cwmheyope, Cwmheiob SO 2174

Cwm Heyop 1815, Cwm Heyhope 1833

W *cwm* as above with the parish-n. Heyope.

Cwmllechwedd SO 1375

Coumlechwedd 1581, Comlechweth 1607, Comllawell 1670, Cwmllechwedd 1833

W *cwm* and *llechwedd* 'hillside, slope'.

Cwmteuddwr SN 9667

Comot Duder 1332, Comot Deudor late *12thC, Commatheewtheur 1460, K.dayddwr c.1566, Cwmtoythur 1406, Launsanfret by Raeyrdyr 1368, Llansanfrid in Comytoyther 1536-9*

The last two sps. above refer to the village and parish of *Llansanffraid Cwmteuddwr* (misspelt *Llansantffraid Cwmdauddwr* on OS maps) meaning 'church of St Bride in Cwmteuddwr' (compare Llansanffraid-yn-Elfael below). The misspelling of *Cwmteuddwr* seems to have started with the assumption that the n. contained W *cwm, dau* 'two' and *dŵr* 'water' but we actually have W *cwmwd* 'commote, subdivision of a *cantref* or hundred' with *deuddwr* (*dau* and *dŵr*) meaning 'commote between the two waters (the rs. Elan and Wye)'. There is a similar example in *Deuddwr* MTG, the n. of the old commote between the rs. Efyrnwy and Severn north of Welshpool.

Cymaron, Cwm Aran SO 1165

(castle) *Kaminarum 1215,* (court) *Caumaron, Kamaron 1297, Cammaron 1304, Cymaron 1262, castell Kymaron 1144, Castle Combe Arran 1734*

The n. of a medieval castle and a r., wrongly called *Aran* on OS maps through misinterpreting corrupt, wrongly-divided spellings. The river is *auon Kymaron* c.1400 and its n. could mean 'battling river', with W *cymar* 'battle, conflict' and the ending *-on* (as in Ieithon).

Ddyle SO 0575

Ddela 1820-1, Ddyle 1833

Possibly 'black, dark place', with W *du* 'black' and *lle* 'place'. W *tyle* 'hill, ascent' is also possible since several ns. in RAD are spelt *Dd-* but pronounced *D-*, ie. *tyle* might be found as *Dyle* through soft mutation after a preposition such as *i* 'to' or *o* 'from'. Unfortunately, *tyle* seems to be a rare n. in RAD.

Diserth SO 0358

Dysserch c.1291, Dyssard 1291, Disserth 1353, y ddiserth c.1566, Thisserth 1554

W *diserth* comes from Latin *disertum* and meant 'hermitage, asylum, refuge' as well as 'wilderness'. Indeed, Diserth is a deserted place; there is no village around the church, only one house (*Red Lion*, a former inn) and a few caravans. The likelihood is that the church was built on the site of a hermitage or in a wild, uncultivated place. The church stands on the western edge of its parish beside the river Ieithon. There is no need to spell the n. with *-ss-* (as the OS). There are like pns. in FLI (misspelt *Dysserth*) and MTG (misspelt *Disserth*); the last occurs as *Disserth* 1309 and *Ddisserth* c.1606 but there is no record of any hermitage or church there.

Disgoed, Discoed SO 2764

Discote 1086, Dichcote 1329, Discot 1563, dishcote 1588, Discoide alias Ditchcote 1591-2, Iscoyde 1598

'Cottage or hut by Offa's Dyke', with OE *dīc* and *cot(e)*. The n. has taken on a W appearance through association with *oddi is*, *'dis* 'under' and *coed* 'trees, wood'. Offa's Dyke passes on the western side of Discoed.

Dolau SO 0266

Doley 1563, 1815, Dolau, -uchaf 1820-1, Upper -, Lower Dolly 1822, Dole Dylase 1716

'Water-meadows (by the r. Dulas)'. The village developed around the Baptist chapel (built 1767) and *Dolau Isaf* (*isaf* 'lower' in contrast to *uchaf* 'upper') farm. The n. is sometimes confused with the *Dolau* near Llanfihangel Rhydieithon. This is called *Doley Gwenffroid* 1563, *the Dolley* 1575, 1753 and *Dolau* 1820-1 and 1833, and applies properly to the houses near the railway station and chapel (and the stream formerly called *Gwenffrwd*) but the n. has become attached to Llanfihangel Rhydieithon. Life must have been rather confusing in RAD: we have another *Dolly* (so spelt 1833), also pronounced *Doli*, near Whitton, which gave its n. to *Dolley Green*. This is called *the Dolye* 1564 and *Dol'y* 1815.

Doldowlod SN 9962

Dol y Daflod, Dol dowlard 1820-1, Doldowlod 1833, Doldowlad 1798

Probably 'loft meadow', with W *dôl* as in Dolau and *taflod, tawlod*.

Dolley Green see **Dolau**

Dolycannau SO 2049

Dole y Canney 1742, Doley-canney 1833, Dol-cannaid c.1817

Probably 'white meadows', with W *dolau* as in Dolau and perhaps *cannaid* 'bright, white'; the second el. also occurs as a rn., e.g. at *Abercannaid* GLA. *Dolycannau* is near a small, unnamed tributary of the r. Arrow.

Dolyhir SO 2457

the Dole-hir Lime Kilns 1833

Either W *dôl* or *dolau* (see Dolau) and *hir* 'long' meaning 'long meadow' or 'long meadows'. W *hir* has a pl. form *hirion* but can occur in its sg. form with a pl. noun.

Drewern SO 1356

Drawen 15thC, 1833, Drewen 1613, y Drewen 1711, Trewerne 1649

Probably a rn. of obscure meaning with parallels on the European continent (*Drava* etc.) according to R.J. Thomas though it could have the same root as W *drewi* 'to stink' and the ending *-an*. *Drewan* might produce *Drawen* by 'vowel inversion'. Popular etymology has produced *Drewern* by associating it with *Trewern, Drewern* – see Trewern below. 'Stinking river' is not that unlikely; both Llandrindod and Llanwrtyd BRE had 'stinking wells'.

Drysgol SN 9473

Droscol 1833

There are similar ns. near Brynaman CRM and Llanymawddwy MER, both referring to hills, and this *Drysgol* too seems to take its n. from an adjoining hill (*Troscol* 1820-1). They all appear to mean 'rough place', containing W *dyrys*, earlier *drys* 'wild, rough', found in words such as *drysgoed* 'thicket', and is essentially the same word as *trwsgl* 'clumsy'. Drysgol has the ruins of a small building – thought to have been an ancient chapel – near Mynydd Clas in the wild, nearly inaccessible valley of afon Marcheini Fawr, north-west of Saint Harmon.

afon Dulas

Aberdulas 1280, Diulas 1241, Dulas flud 1610, Dulas River 1833

W *du* 'black, dark' and *glais* 'stream'. Dulas may be a double-headed stream: one feeder is *Dulas*, the other *Black Brook* – probably an E translation of *Dulas*. The combined stream flows past Nantmel to the afon Ieithon just above Llandrindod.

Dulas is a very common rn. in Wales. In neighbouring BRE we have *Dulas* rising on Gorllwyn and running to the r. Irfon. This is *Delas* 1547, *Dylas flu.* 1578, *Tir blaen Deylus* (near its source) 1619. Nearby is the *Dulas* which rises at the western end of Mynydd Epynt, flows through Tirabad, and reaches Irfon at *Aberdulas*. This is *Dulais* 1833, *Dulas Brook* 1828.

Duthlas SO 2077

Dothless 1814, Duthlis 1828, Dothlass 1862

Without earlier sps. it is too hazardous to suggest any meaning but it could be a rn., almost certainly W. The combination of letters *-thl-* looks like an attempt to represent W *-ll-* (see Ackhill above).

Dyffryn Elan see **Elan**

Dyffryn Gwy SN 9471

Dyffringowy 1544, Diffryn gwy 1597, Dyffryn Gwy 1677, Diffrin Guy 1670

One of two townships in Cwmteuddwr parish adjoining the afon *Gwy* or *Wye;* the other is *Dyffryn Elan* (see Elan below).

Ednol SO 2364

Edenwalle 1342, Adenwale 1397, Ednole 1549, Ednoll 1587, Ednall 1544, 1670

OE, probably a pers.n. *Ēada* and *wall* 'wall or bank' or *well, wælle* 'well, spring, stream'. Now only a farm and a few houses but there was a chapel-of-ease (of Old Radnor) and the ruins remain.

river Elan

Glan Elan 1184, Eleyn 1241-50, Aberelan 1581, Alen, Alan 1536-9, Driffyn Elan 1426, Duffrin Elan 1688

Probably W *elain* 'hind, fawn'; animals-ns. are often applied to rs., no doubt in allusion to their nature. Before Birmingham Corporation built the reservoirs and dams here in 1892-1904 the lower course of the Elan must have rushed and leaped over the rocks like a hind.

The mountainous area at the head of the Elan, Severn, Teifi, Wye and Tywi seems to have been called *Elennydd* and it is recorded as *Elennith, -yth* c.1200, *Elennith* 1536-9 and *Elenid* 1446-89, meaning simply 'area adjoining the river Elan'. The ending *-ydd* often had the sense of 'territory' in Wales as in *Meirionnydd* 'land of Meirion' and Maelienydd. The township around the Elan was called Dyffryn Elan, 'Elan valley'.

The flooding of the valley involved the construction of four large dams and the destruction of the old chapel Capel Nant-gwyllt and two important mansions Cwmelan and Nant-gwyllt, well-known for their associations with the poet Shelley. From bottom to top the reservoirs are *Caban-coch, Y Garreg-ddu, Penygarreg* and *Craig-goch* named after three small farms given as *Caban-coch* ('red hut'), *Carreg-ddu* ('black rock') and *Pen-y-garreg* ('top of the rock') and after the hill *Craig yr allt goch* ('crag of the red slope') 1833.

Elan rhymes with 'hell an' (high water)' not E 'eel'.

Elfael

Eluen 1142, Elevein 1198, Eluail, elmail 12thC, Elwael c.1100-3, Elvael 1267-76

Probably a W pers.n containing *El-* as in the old ns. *Elci* and *Elgnou* with *-mael* as in Nantmel. The form *Elwell* often seen on maps until recently is likely to be an E effort at sp. W *Elwel*, a variation of *Elfael* (for a similar variation of W *w* and *f* see Doldowlod).

Elfael was an ancient *cantref* divided into two commotes *Uwch Mynydd* (*Eluael uch Mynyth* 1276) and *Is Mynydd* (*Elvellismenyth* c.1277) meaning 'above the mountain' and 'below the mountain', otherwise known as Colwyn and Painscastle hundreds from their main castles. Between them lies the highland stretching from Llandeilo Hill to Glascwm Hill. Elfael covered the southern part of RAD.

Esgairnantau SO 1762

Esgair-nantau 1820-1, 1833

'Ridge characterised by streams', with W *esgair* 'ridge' and a pl. form of *nant*. Earlier spellings are needed for certainty.

Evenjobb : Einsiob SO 2662

Emynghop 1304, Emyngehope 1328, Evinyheope, Evyngeopp 1544, Evengobb 1653, Evengeobbe 1612, Angob 1670, Einsiop 16thC

'Enclosed valley associated with a person called Emma', i.e. *Emma* and *-ingehope*. The change from *Em-* to *Ev-* may be a result of W influence comparable with the soft mutation *m* to *f* (a *v* sound) commonly found in W. The pn. then contracted in the same way as Burlingjobb to *Burchop*. The W form *Einsiob* is not well attested but seems to show a further contraction comparable with the E adverb *even* to dialect *e'en*.

Faenor SO 0169

'r Faenor 16thC, Vaynor 1544, VAYNOR 1688, Great -, Little Vaynor 1820-1

Properly *Y Faenor* with the def.art. W *maenor* is fem. and fem. words beginning with *m* soft mutate after the def.art. *Maenor* can be loosely translated 'manor' but is unlikely to be a borrowing from the E word. It seems to have meant originally 'a chieftain's residence with a wall around it', perhaps because *maenor* seems to contain W *maen* 'stone, boulder'. It later came to mean 'group of villein hamlets' and 'subdivision of a *cantref*' under the influence of E *manor*. As a pn. it is sometimes applied to a particular township – perhaps this is where the lord's house lay in Nantmel parish? The farm *Little Vaynor* is probably *Tythin y Vaynor alias Cwmcian* 1719, i.e. W *tyddyn* 'cottage, smallholding' while *Cwmcian* may be W *cwm* and *cuan*, 'owl's valley'.

Faenor-glas SO 1352

o'r Vaenawr 1446-87, veynor Glascun c.1290-93, Veynor glare 1623, 1671, VAYNOR-GLARE 1804, Faynor c.1817

W *maenor* as in Faenor above and an abbreviated form of Glasgwm. Some sps. have *glare, Glair* etc. which has not been explained: is it a rn. with *claer* as in Claerwen?

Farrington SO 3070

Farington 1670, Farringdon 1688, Farrington alias Farneton 1790

The sps. are too late to be reliable. We may have OE *fearn* 'fern' with *tūn* or *dūn* meaning either 'fern settlement' or 'fern hill'.

Felindre SO 1781

Vellendry 1695, y Velindre c.1700, Felindre 1832

'The mill township', with the W (lost) def.art., *melin* 'mill' and *tref* now usually meaning 'town' but earlier 'township, hamlet' and even 'farm' (like OE *tūn*). *Felindre* (SO 1836) BRE is *Capel y felindra, Pentre y felindre* c.1700), *Felindre* 1832, *Velindre* 1744. Both villages have suitable mill-streams. D. Silvan Evans, a W scholar, once objected that there were '*melindres* where no mill ever existed' and sought a dubious derivation from '*milein-dref* = the dwelling of the *milain* or *villein* of feudal times' but ignored the possibility that the 'mill' might be a horse-mill or a windmill (rather than a water-mill) and that the 'mill township' might simply mean a township which was expected to use the manorial mill. The evidence for the use of W *milain, bilain* 'villein' in pns. is very scanty.

Franksbridge: Pont-ffranc SO 1156

pont ffrank, Frank bridge c.1700, Frank's Bridge 1746,1786

Probably 'bridge associated with a person called Frank' but the sps. are too late to show whether the pn. was originally W or E. If it is E then we may have E *frank* in the sense 'free', i.e. 'bridge which did not pay tolls', but that was true of many bridges before the late 18thC. W *ffranc* 'mercenary' is also possible and has been suggested for *Coed-ffranc* near Swansea GLA but *ffranc* in pns. needs further investigation since it sometimes occurs with tree-ns. and *rhiw* 'slope'. Apart from *Coed-ffranc,* note *Llwynfranc* (Llanfihangel Crucornau) MON, which occurs in *Great -, Little Llwyn Frank* 1832; also *Penrhiw Frank* (*Penrhiwfranc alias Brocho* 1833), *Troedrhiwfranc* and *Coedrhiwfranc* (Cefnllys parish) RAD and there was a parcel of land called *Y franc wydd* 1624 (Bachaethlon, MTG).

Fron SO 0865

No. sps. seen; possibly W *bron* 'hill-breast, hill-side' applying to the hill on the east side of the A483.

Gamallt hill SN 9570

Kamall c.1755, Gamallt 1820-1, 1833

'Crooked slope', with W *cam* 'crooked, bent' and *allt* 'slope, wooded slope'. The change from *Cam-* to *Gam-* may be the effect of a lost def.art. (*Y Gamallt)* because *allt* is fem. or the effect of a preposition such as *i* or *o (i Gamallt, o Gamallt).*

Garreg-ddu Reservoir see **Elan**

Gladestry : Llanfair Llwythyfnwg SO 2355

Lanfeyr Lonthonnok c.1291, ll.fair llwyth ynnoc c.1566, Llandouenoge 1566, Claudestre c.1250, 1304, Glandestre 1334-5, Glawdestrey 1517, Glawstree 1522

OE *glad, glæd* 'glad' occurs in the surnames *Gladwin, Gladden* and may have produced a pers.n.; the second part is likely to be *trēo(w)* 'tree, cross' (as in *Oswestry, Coventry* etc.) and the ns. of a number of hundreds in England, probably originating as the ns. of their particular meeting-places.

The W n. is just as intriguing, though its spellings vary considerably. *Llanfair* is W *llan* 'church, church enclosure' and *Mair* 'Mary' to whom the church is dedicated. The remainder is the n. of the district which occurs independently as *Comm. Loethifnuc* 1536-9, *llech ddyfnoc* 1543, and the poet Lewys Glyn Cothi (1446-89) makes several mentions of it, such as *dir llwythofnwc*. Its correct spelling is a little uncertain but it appears to be composed of W *llwyth* 'tribe, clan' (or its area) and a pers.n., perhaps *Dyfnog* (though Payne preferred *Yfnog)*; i.e. *Llwyth Ddyfnog* would become *Llwythyfnog, Llythynnog* etc. The 1566 spelling seems to represent modern W *Llanddyfnog*.

Glasbury : Y Clas-ar-Wy SO 1739

Clastbyrig 1056, Glasbir' c.1185-90, Glasbur' c.1291, Glasburia 1144, Claysbury 1460, a'r clas 1447-86, y Klas ar Wy c.1566

Local people call it 'Glazebury' and it is possible that the pronunciation has been influenced by OE *glæs* 'clear, bright, shining' but it appears to be a hybrid of W *clas* (see Clas Garmon) meaning 'monastic community' with OE *burh* (dative *byrig*) usually indicating 'fort, enclosure'. No fort has been found here and it is possible that the *burh* refers to an enclosure centred on the *clas,* wherever that was. W speakers in the Erwood BRE area 100 years ago called Glasbury *Y Clas* and there was a wood *Coed y Class* recorded in the parish in 1760. The second part of the W n. means 'on Wye'.

The old church stood between the rs. Wye and Llynfi in the *Stances* or *Stonces Meadow* very close to the modern village but was demolished in the 1660s because of repeated flooding. The new church was built on the BRE side on ground called *Clas* (or ? *Clos* meaning 'close') *dan Coed y Bolin* and consecrated in 1665. The RAD part of the parish obtained its own church All Saints 1881-2. *Coed y Bolin* may have W *polyn* 'pole' although it is usually a masc. word and one would not expect the mutation of *p-* to *b-* found in fem. words after the def.art. Perhaps *y Bolin* is poor spelling of W *ebolion* 'colts'.

Glasgwm, Glascwm SO1553

Glascum 1188, glasgwm c.1180, Glascom 1215-22, Glascomb 1407, LANDOU GLASCOM IN ELUAEL 1292

Probably the colour *glas* 'blue', rather than *clas* as in Glasbury, with *cwm* meaning 'blue, green valley' although the scholar Edward Lhuyd c.1700 mentions 'Klas . . . a small lordship' belonging to the Bishop of St David's here and the stream running through the village towards Cregrina is *Clas Brook*. The last, however, is likely to be a 'back-formation' from *Glasgwm* and *Klas* must refer to Faenor-glas. *Glas* is a very common pn. element hereabouts: there is a stream (and formerly a farm) on the eastern edge of the parish called *Glasgwmig (Glascwm-ig* 1833) meaning 'little Glasgwm' and a farm *Glasnant* (so spelt c.1817) at SO 181509. *LANDOU* 1292 presumably refers to the church dedicated to *Dewi* (David) i.e. *Llanddewi.*

Glog Hill hill SO 2269
Glog Hill 1833

W *clog* 'crag, cliff' though in 1815 it is *Fron Hill;* compare Fron above.

Golon approx. SO 0771
Gollen 1541-2, 1566, Golon 1535, 1689, Gollon 1769, Colland 1670

Unexplained and there seem to be no clues in the topography to its meaning. It is just possible that *Golon* has developed from *collen (y gollen* with the def.art). 'hazel-tree' since instances do occur where W *ll* has become *l,* not simply in areas where E replaced W at an early date.

Gorddwr SO 1581
Gorddwr c.1600, Gwr-dwr Mill, Cwrd-dwr Bank 1836, Gorther 1828

That area 'beyond the water' or simply 'upper reaches (of the r. Teme)', containing W *gor-* (also *gwor-, gwar-)* and *dŵr*. Also the n. of a lordship in MTG *(le Cordur* 1277, *Gortwor 1278, Gordor', Gordowyr 1265, Overgorther, Nethergorther 1305)* on the eastern side of the r. Severn around Buttington, having 'upper' (*over*) and 'lower' (*nether)* parts.

Gorslydan SO 1276
Gors ludan 1833

'Wide bog', with W *cors* 'bog, fen' and *llydan* 'wide, broad'.

Gumma SO 2965

Gommey 1550, Combeya 1585, Gomey 1553, Gummey 1606

This is likely to be the source of the surname *Gomey, Gummy, Agommey* and *a Gombe* which occurs at Presteigne 1506-68. Possibly W *cymau* 'valleys', pl. of *cwm* or perhaps OE and ME *comb* 'crest' and *ēg* 'dry ground in a marsh' though the n. now applies to a hillside farm.

Gwaithla: Gweilliwr SO 2156

Gwiler, -or 1670, Gwillour 1675, the Gwillour 1693, Gwyllwr 1692, y Gweillwr Ros, o'r Gweillwr c.1700, the Gwiller 1738, Gwaethellan 1598, Gwilther 1804, Gwithla 1833

Possibly W *gwilliwr* 'wanderer' referring to Gilwern Brook or the small stream on the south side of Gwaithla Farm since Gilwern Brook was also called Weythel. Despite the rough similarity of sp. *Gwaithla* and *Gweilliwr* do not appear to have any connection with *Weythel*. The modern sp. *Gwaithla* is fairly recent, a product of E speakers replacing W *-ll-* with *-thl-* and then 'obscuring' the vowel and *-r*.

Gwastedyn SN 9866

Gwestedyn vawr 1614, Gwastedyn Vawre 1657, (hill) *Gwastedin 1695,* (hill) *Gwasteden 1820-1, Westedin 1544*

A former township in Nantmel parish dominated by *Gwastedyn Hill*. The n. may derive from the hill, perhaps meaning 'small flat place' (W *gwastad* 'level, flat' and a diminutive ending *-yn*), though the hill is not very flat on top and comes to two slight points at *Carn Fach* and *Carn Wen* ('little cairn, white cairn').

Gwaunceste Hill hill SO 1555

Gwaun Cestu c.1817, Gwaun Ceste 1833

The farm at SO 176548 is *Cwm Cesty* c.1817, *Cwm-ceste* 1833 and there is a *Kesty* on modern maps at SO 179541. *Cesty* or *Ceste* might be W *cais* 'seeker, inquisitor' (found in medieval records as *keys*) rhyming with E 'twice' and *tŷ* 'house', i.e. *ceisty* meaning 'house of the bailiff or inquisitor' or something very similar. *Cais* has become *Ceis-* because of the effect of the vowel on *tŷ* ('vowel affection').

Gwerthrynion

Warthrenniaun c.1200, Warthreynnon 1304, Gwerthrynnyaun 1276, gwarthrynion c.1560-65

'Land of Gwrtheyrn', i.e. the W pers.n. and the ending found in, for example, *Ceredigion* ('land of Ceredig'). Some of the consonants and the vowel *-e-* have become mixed (*-theyrn-* to *-thryn-*) by a process called 'metathesis' (compare 'Brummagem' the old pronunciation of *Birmingham*). Nothing definite is known of Gwrtheyrn but his n. is identical to that of the legendary *Vortigern* (OW *Guarthigirn*, later *Gwrtheyrn*) in Nennius's *History of the Britons*; the n. seems to mean 'great ruler'. Vortigern was notorious for his wickedness and is said to have fled from the saintly Germanus or Garmon (see Saint Harmon).

Gwystre SO 0558

the Goostry 1796, the Goosetree 1822, Goosetree 1822, Gwystre 1828, 1833

Probably 'sow farm or settlement', with W *gŵys* and *tref* (see Felindre) and comparable with *Mochdre* ('pigs farm') of which there are examples in MTG and DNB. The sp. ought to be *Gwysdref* but hardening (or 'provection') has turned *-d-* to *-t-* – common after *-s-* – and final *-f* is lost in spoken W. Local people call it 'Gwstri' as shown by the first three sps. cited above and it is possible that after this area became anglicised in the late 18th and early 19thC the pn. was misunderstood as having some connection with E *goosetree*, explained in the *Oxford English Dictionary* as 'the tree from which barnacle-geese were believed to be produced'.

Harpton : Tre'rdelyn SO 2359

Ortune, Hertune 1086, Hortona 1137-9, Harpeton 1304, Harpton 1343, Trer delyn c.1541, Tref y delyn c.1569

Both the ModE and the W n. are misleading. The earliest sps. suggest it is composed of OE *horh, horu* 'filth, shit, slime' and *tūn*; at best it might mean 'settlement in muddy land'! ME *harp* seems to have been substituted before 1304 to clean up the pn. (if not the place) and W *telyn* 'harp' (with prefixed def.art.) translates the new E word; the substitution may not have been deliberate but a matter of simple confusion.

Hen-goed SO 2253

Hengoid 1649, Hengoydd 1688, Hen-got c.1755, Hencoed 1828

'Old wood or trees', with W *hen* 'old' and *coed*; there is another *Hengoed* and a *Burnt Hengoed* (both farms) in the adjoining parish of Huntington HER.

Hergest Ridge hill SO 2556

Hergest, -h 1086, Heregast 1251, 1373, g6rt herest, herest, hergest lys, argl6yd Herast 1447-89, Cwrt Hergest c.1420-1500, hergest 15thC, (lord of) *Herest, Hergest 1600-7, HERGEST RIDGE 1815, Hargast 1828*

Uncertain; for *Hergest* HER – just over the border – see Coplestone-Crow 116. The second el. could be W *cest* 'belly' in a topographical sense like *Borth-y-gest* CRN (with W *porth* in the sense of 'harbour') but the first el. is very doubtful; W *hir* 'long' seems unlikely because of the sps. *Her-, Har-*. The forms with *lys* and *cwrt* mean much the same, i.e. W *llys* and *cwrt* 'court, mansion'; *argl6yd* is ModW *arglwydd* 'lord'.

Heyope : Heiob SO 2474

Llandewy in Heyop 1389, Heyope 1398, Heyhop 1535, Heiop 1692

'High valley', with OE *hēah* and *hop*. The church is dedicated to St David (*Dewi*), hence the 1389 sp. for *Llanddewi* (see Llanddewi Ystradenni). Near the top of the valley is

Cwmheyope and the site of a former house called *Tuy yn Mlaen Heyop* 1672 ('house in the upper reaches of Heyope').

Hirddywel hill SO 0280

hirddywel c.1268, (forest) *Hirthowel 1356-7,* (mountain) *Herdoel, Heerthowell 1587, Herethowell 1382, RHYDD HYWEL 1833*

The n. is now applied to the hilly area east of Pegwn-mawr and Pegwn-bach in Llandinam MTG but formerly described a broader area extending into Ceri (Kerry) and RAD. Until recently OS maps misspelt the pn. *Rhyddhywel*. The meaning is not clear but it may be W *hir* 'long' and perhaps *dywal* 'fierce, cruel' (this is found in field-ns. elsewhere in Wales); the sense may be 'long wild (hill)'.

Howey : Hawy SO 0558

(grange) *Blahenhavoe 1297,* (r.) *Hawye 1578, Rhos havoy 1588,* (mill) *Havy 1591-2, Melin hafwy ycha 1679, Howy 1828*

The village of Howey is clustered around a small bridge over the r. *Howey*. The correct sp. of both is probably *Hawy* or *Hafwy* meaning 'summer stream', i.e. W *haf* 'summer' and *-wy* commonly found in rns. The first sp. above has W *blaen* 'source or upper reaches (of a r.)'; the third has W *rhos* 'moor' and the fifth has W *melin* 'mill' with *uchaf* 'upper'.

Hundred House SO 1154

Hundred Ho. c.1817, Hundred House 1833

The former meeting-place of the hundred of Colwyn. There is another *Hundred House* in Bleddfa, perhaps a meeting-place in Maelienydd.

River Ieithon, Ithon

yeithyon 12thC, Yeythan 1297, ieithawn 1447-89, Ithon 1591-2, ython 1580

W *iaith* with the ending *-on* meaning 'talking river' from its noise and babbling. Local people pronounce the n. 'Eithon' (rhyming with E *python*). The tendency to drop *i* (when it acts as a consonant) is found elsewhere in RAD and Wales.

Iscoed

Hiscoyd 1338, Iscoyd 1360-2, Ishcoed 1688

W *is* and *coed* meaning 'below the trees or wood', used in other parts of Wales for the lower portion of a *cantref* (in this case Gwerthrynion) around the lord's residence and demesne (*maerdref*); contrast Uwchcoed for the upper portion. Iscoed was the south part of Gwerthrynion including Llan-ŷr, Llanfihangel Helygen and parts of Nantmel.

Kinnerton SO 2463

Kynardton 1304, Kynarton, Kinarton 1343, Keynarthe alias Keynarton 1549, Kinart 1670, Kinert cap(el) c.1700

OE 'Cyneheard's settlement' and identical to *Kinnerton* FLI. The W n. seems to have been *Cinart* or something very similar and shows the loss of the final syllable *(-ton)* like Barland and a number of pns. in DNB and FLI.

Knighton : Trefyclo SO 2872

Chenistetune 1086, Cnichteton 1191, Knitteton 1207, Knyghteton 1230, Knighton in Walche . . . Trebuclo 1536-9, (castle) *Trefeclaud 1262, Trefycclawdd 13thC, Trebuclo pro Treficlaudh 1586*

'Farm or settlement of the servants', with OE *cnihta* pl. of *cniht* 'youth, servant, soldier' and *tūn*. The W n. means 'settlement or farm of (or by) the dyke' referring to Offa's Dyke, with *tref,* def. art. and *clawdd* 'bank' (and 'ditch'). Stress evidently lay not on *Tref-* but on the def. art. (*y*) and as a result *-aw-* was 'rounded' and shortened a little to produce *-o-* and finally *-dd* was dropped in ordinary speech. R.J. Thomas says that *Trefyclo* contracted further to *Tryclo* but I have not found any evidence of this. The castle here is called *C(astell) y Clawdh* late 16thC. *Tref-y-Clawdd* is incorrect.

Knucklas : Cnwclas SO 2574

Cnuclays 1246, Knoclas c.1220-7, Cnoklas 1241, C(astell) y Cnwclas late 16thC, Knucklas 1670

W *cnwc* 'hillock' and *glas* 'blue, green' referring to the green mound of the castle. The initial *K-* (properly *C-*) is no longer pronounced by local people – a result of anglicisation. Both W *cnwc* and E *knock, knuck* (in pns.) may derive from OI *cnocc* 'hill'.

Lanlluest SO1874

Llan-las 1820-1, Lluest Bank 1833

Possibly W *glan* 'bank' and *glas* (as in Knucklas) meaning 'green bank' or *glan* and *lluest* 'hut' (later 'tent') but the pn. needs further investigation.

Litton SO 2666

Gliditon 1304, Glitton 1487, Glyton 1545, Letton 1340, Llitton 1610

OE, probably a pers.n. from *glida, glēoda* 'a kite, a bird-of-prey' and *tūn* 'settlement or farm'. The loss of *G-* may be W influence, i.e. the common soft mutation after a preposition such as *i* 'to' and *o* 'from'.

Llaethdy SO 0680

Laythde 1232, Laith-Ty Ucha, - Issa 1769, Llaeth-du c.1818-21, Llaethdy-uchaf, -isaf 1833

'Dairy', with W *llaeth* 'milk' and *tŷ* 'house' etc.; compare Cornish *Laity* and Breton *Laedti*. The n. is fairly common in Wales as a minor n., e.g. *Llaethdy* near St David's PEM *(Laythte* 13thC) and there is a *Nant y Llaethdy* in Dyffryn Crawnon BRE.

Road-signs and maps persist in misspelling the n. *Llaithddu* which has been interpreted as 'black swamp' but *llaith* is an adjective (not a noun) meaning 'damp, soft'. When Powys County Council erected signs in 1985 showing the correct sp. some local people objected and the *Llaithddu* signs had to be replaced!

Llan see **Llanddewi Ystradenni**

Llananno SO 0974

Thlananno 1304, Lammannowe 1332, Llananno 1392-3

The church is dedicated to *Anno* but nothing reliable is known about him. The old n. of Newborough ANG is reputed to have been *Llananno* but this is doubtful (PNANG 106).

Llanbadarn Fawr SO 0864

Thlanpadervau 1304, Lanbadarn Vaur 1374, Lampadern' c.1291, ll.badarn fawr c.1566, Llanpadern Vare in Melenyth 1386, Lampeter in Melenyth 1401

W *llan* 'church' (earlier 'enclosure' as in *perllan* 'orchard'), the saint's n. *Padarn* and *mawr* 'big, great' referring to the importance of the church or parish in contrast, presumably, to Llanbadarn Fynydd. Llanbadarn Fawr may have been the 'mother church' in this part of Maelienydd. In some sources it is difficult to distinguish the RAD *Llanbadarn Fawr* from the identical pn. in CRD near Aberystwyth.

Llanbadarn Fynydd SO 0974

Llanpaderdyffeyht 1332, Llanpadern 1392-3, LLAMPADARN in Melyny 1406, Llanpater[n]venyth 1431 Llanbadarn Vynith 1513, (cemetery of) *St Padern (Sancte Paterne) in the Desert 1399*

'St Padarn's church in the mountains', with *mynydd* 'mountain', in contrast to Llanbadarn Fawr lower down the Ieithon valley. The alternative qualifying word is W *diffaith* 'poor, unfruitful', rendered 'desert' in one source; see also Maelienydd for the 1406 sp. There is a mention c.1291 of *Lanemeth* which may be this place, possibly an attempt at spelling *Lanemeneth* (= *Llanfynydd)* – a short version of *Llanbadarn Fynydd*?

Llanbadarn-y-garreg SO 1148

Thlanbaden c.1291, ll.badarn garreg c.1566, Llanbadarn y Garregg 1623, Lampeter 1401

Perhaps 'St Padarn's church near a boulder', with W *carreg* 'rock, stone', or elliptically to mean a place in rocky countryside (compare Llanbadarn Fynydd).

Llanbedr Painscastle : Llanbedr Castell-paen SO 1146

Lampetir Pain Castle 1283, Lanpedr', Lampetr' c.1291, Lambed 1559, llanbedr 1447-86, LLANBET 1599, Lambet Castlepaine 1620

'Church of St Peter by Painscastle'. Some sps have *Lam-, Llam-* as in Llanbadarn Fawr, and *Lampeter* (= *Llanbedr Pont Steffan*) CRD, a feature common to both W and E pns. when *-n-* occurs before *-b-*.

Llanbister SO 1073

Lambister 1252, Lanbyster 1334, Thlambister 1297, llann bisteir 1447-89, Llanpister 1401

There is no known 'St Pister' to explain this pn. and the church is dedicated to *Cynllo* as in nearby Llangynllo. The first part must be W *llan* (see Llanbadarn Fawr). There are a few sps. with *-vister* (= *fister*) but the majority have *-bister* implying an original *pister* but the word has not been explained.

Llanddewi Fach SO 1244

llann de6i y crwys 12thC, Llanddewi isvynydd, – Ismynydd 1548, ll.dewi fach c.1566

'Lesser church or parish of St David', perhaps in contrast to Glasgwm or Llanddewi Ystradenni. Once a chapel-of-ease to Llowes. The earliest form has the W def.art. and *crwys* 'cross, crosses' but there does not appear to be any standing cross here to explain it. The church stands in an isolated spot on the southern slopes of *Allt Ddewi* (*Allt-dewi* 1833), 'David's slope' or (better) 'slope by St David's church'. Llanddewi Fach lay in the commote of Ismynydd (see Elfael).

Llanddewi Ystradenni SO 1068

Thlandewy 1297, Llandewy 1332, ll.dewi ystrad yni c.1566, Llanddeywy Stradyny 1582, – ystrad nynhid, ystrat nynhit 12thC

'Church of St David in the vale of Nynnid'. *Nynnid* may be the fem. pers.n. which was latinised as *Nonnita* but there is evidence also of a masc.n. This may occur in *Eglwys Nynnid* GLA and *Hendrenenig* DNB. Local pronunciation is 'Lan-jŵi, and road-signs ignore the ancient *Ystradenni*. That is ironic since *Llanddewi* may be the later n., added to *Ystradenni*. The township in which the village stands was called *Llan* or *Tre'r-llan* (*Tre yr llan* 1635), the second meaning 'the church township', with *tref* (as in *Trefyclo*; see Knighton).

Llandegle SO 1462

Landegle c.1291, Llandegle 1401, Llandegley 1513

The church is dedicated to *Tegla* or *Tegle* as in the case of *Llandegla* DNB. Most locals now call it 'Landegli'. The nearby moorland *Llandegle Rhos* is called *Llandeglaye's -, Llandeglie Rhose* 1557; *rhos* 'moor' was borrowed from W into the E dialect of RAD.

Llandeilo Graban SO 0944

Lann teliau iciliou idifrinn machagui 12thC, Llandeylar' c.1291, Llanteylo in Elveal 1397, ll.deilio c.1566, Llandilo yr Graban 1688

Llandeilo is easily explained as W *llan* (see Llanbadarn Fawr) and *Teilo*, the n. of a saint recorded before AD 700 in the Book of St Chad. Graban seems to be a late addition possibly taken from nearby *Cwrt-y-graban*; this is W *cwrt* in its sense of 'mansion', the def.art. and perhaps *graban* 'corn-marigold', but doubts linger: Cwrt-y-graban is two miles from Llandeilo church and lies in Llansteffan parish.

In the first reference Llandeilo is associated with *Ciliau* which stands above the junction of the r. *Bachawy (machagui)* – misspelt *Bach Howey* on OS maps – with the Wye. *Ciliau* is the pl. of *cil* 'nook' (see Cil-gu). *Bachawy* is an interesting n.: *machave, machavvi, Drifrinn Machagui* 12thC, *machawy* 1447-86, *Glanmahawye* 1549, *Machewoid, Machtawoyd* 1578, probably deriving from a pers.n. *Machafwy, Machawwy* (according to R.J. Thomas). The initial *M-* has become *B-* partly because there is a tendency in W to confuse them (they have similar mutations) and partly because *Machawy* may have been associated with common W words such as *bach* 'little' and *bachog* 'hooked'.

Llandrindod Wells SO 0660

LLANDYNDDOD 1535, Llandrindod 1549, Llan Yr dryndot 1543, Llanydrindod 1565, Llandrindod Wells 1794

The old church dedicated to the Holy Trinity (W *trindod*) stands on a hill overlooking the modern town which lies partly on a former common. The modern church in the town was erected only in 1871 and the town owes its origin to the popularity of its mineral springs, hence the addition *Wells*. Some early 19thC writers simply call the place *The Wells*.

The old church was earlier called *Llandow* 1283, *Lando* c.1291, *Landon* (= *Landou*) 1353, meaning 'church dedicated to God', i.e. *llan* and *Dwyw* (ModW *Duw)* 'God', like *Llanddew* near Brecon. Some churches in Wales with similar dedications were re-dedicated to the Holy Trinity in the late Middle Ages – after all, are not all churches dedicated in a general sense to God? Llandrindod once possessed another church, excavated by archaeologists in 1984, called *Llandevailow* 1283, *Lonvayloir, Landemaylon* c.1291, *Landevalow* 1331, *Landevaelon* 1353, *Llanvaylon* 1551, *Llanvello in Llandryndode* 1559. The n. survived as late as 1732 as an alternative n. *Llanvailon* for *Glanbongam Isaf* (now *Llwynbongam*) farm (according to R.C.B. Oliver). The correct modern sp. is uncertain but it is possible that we should use *Llandyfaelo* or *Llandyfaelyw* since the final *-n* in some spellings may be copyists' errors for *-u*. No saint *Tyfaelo, Tyfaelon* or *Tyfaelyw* is known but the pers.n is very similar to *Tyfaelog* seen in *Llandyfaelog* CRM and *Llandyfaelog Fach* BRE.

The basis of both ns. is the honorific *ty* (W for 'thy') with *mael* (as in Nantmel) but it is not known whether the saints are identical. The foundations of the excavated church have been re-erected near the Lake and the notice-board describes them as 'Capel Maelog', a n. first used apparently by Jonathan Williams in his *History of Radnorshire* c.1818. The only other support for this sp., seems to be *Cartre Faelog*, a field near the original site of the church close to Cefnllys Lane.

Llanelwedd SO 0451

Llanelweth 1283, Lannelwyth 1334, LL.elwedd c.1560-66, Thannelwyt 1291

'Church of St Elwedd', with W *llan*. Lewys Glyn Cothi (1447-89) the poet records the existence of a saint of this n. in his *cywydd* to the saints of Elfael and Llanelwedd certainly lay within the boundaries of Elfael. *Elwedd* may be the same as *Eiliwedd*, one of the legendary twenty-five daughters of Brychan Brycheiniog (from whom Brycheiniog – later anglicised as Brecknock – is reputedly named) recorded in W manuscripts c.1200. The church is now dedicated to St Mathew.

Llanfadog SN 9465

Llanvadauc 12thC, (tenement) *llanvadog issa 1579* (farm) *Nant Madocke 1580*

Presumably 'church of Madog', though *Madog* is not necessarily a saint since the stream here is *Nant Madog*, 'Madog's stream'. The evidence for a church or chapel-of-ease here is very slim; ruins were said to survive here in the early 19thC in *Llwyn y Capel* ('the chapel grove') opposite the Elan Valley Hotel.

Llanfair Trellwydion SO 0973

(messuage) *tre Llwydon 1613, Llanver Tre Llwydon c.1755, Trellwydion 1833*

The site of a reputed chapel, a little south-east of *Trellwydion* farm, presumably dedicated to St Mary (W *Mair*). The remainder of the n. may be 'settlement' (or farm) of a person called Llwydon', with W *tref* (see Felindre).

Llanfaredd SO 0750

Llanvareth 1557, ll.varaith c.1566, Llanwareth 1598-9, Llanverreth 1559, Thlannarreyt 1291

'Church of Maredd (or Mareth)', with W *llan. Maredd* may be identical in meaning to W *maredd* 'pomp, splendour' though some have seen it as a W form of *Margaret.* The small stream on the south side of the church is called *Varedd* 1833 and it is possible that this gave rise to *-faredd* rather than a pers.n. but the stream-n. is not otherwise authenticated and it is likelier that *Maredd* or *Mareth* is identical to the saint called *Mariaith* by Lewys Glyn Cothi in his *cywydd* to the saints of Elfael (see Llanelwedd). The church is now dedicated to Mary *(Mair),* possibly a false dedication inspired by the similarity of *Maredd* or *Mareth* and *Mair.*

Llanfihangel Helygen SO 0464

ll.V'el fach c.1566, Llangyhangell Religen 1559, Llanvihangel Helegen 1587, Llanvihangel-fach Helygon 1832

Llanfihangel means 'St Michael's church' and W *helygen* is 'a willow-tree'. It may seem an odd addition but has probably been applied elliptically to indicate 'an area where willow-trees grow, a wet area or an area close to a stream'. The church is certainly close to *Pentre Brook.* The other addition is W *bach* 'little' (mutated as *Fach*), probably to distinguish this place from *Llanfihangel Cefnllys* (see Cefnllys), or perhaps because the building is small and was a chapel-of-ease to Nantmel.

Llanfihangel Nant Melan SO 1858

St Michael of Nantmelan 1231, St Michael, Llanmelan 1398, o blwyf nant melan 1447-89, ll.V'el nan melan c.1566

'St Michael's church by Nant Melan'. The r. here is called *Avon Melan* c.1700 by Edward Lhuyd who also records a lake *Lhyn maes melan* (W *llyn* 'lake', *maes* 'open field, field'), probably near *Maesmelan* farm (SO 192588). *Melan* may contain W *mael* 'prince, lord' like Nantmel and is an alternative n. for *Summergill Brook: Somergelde 1337, Somergil* 1536-9, 1602, i.e. 'brook dry in summer', with OE *gelde* 'barren'.

Llanfihangel Rhydieithon SO 1566

Langmiclen c.1291, Thlan Mihangel Redyethan 1304, Lanyangel Ytheyan 1332, ll.V'el Rydeithion c.1566, llanvyhangell Rudythone 1549

As Llanfihangel Helygen, with 'ford of river Ieithon': W *rhyd* 'ford'. The odd thing is that the village is two miles away from the Ieithon (Ithon). The township here is called *Rhydython* 1563, *Rhydithon* 1587 so there seems to be little doubt about its meaning. The possibility that the Ieithon was a double-headed r. like Efyrnwy (Vyrnwy) MTG is unlikely because the main r. next to Llanfihangel is called Cymaron and there is no evidence that it was called anything else. Either *Rhydieithon* once applied to a larger area reaching to the r. Ieithon or the n. applied to a ford in the small stream, now called *Maes Brook*, which rises on Radnor Forest and flows through *Cwm Merwys* (W *merwys* 'blackbird') and Llanfihangel to the river Cymaron.

Llangynllo SO 2171

Llankenlowe 1368, -kenllowe 1395, LLANGUNLLO 1482, Llangenllo 1407, llann gynllo 15thC, Llangetlau 1323

'Church of St Cynllo'. Little is known about him but he was the leading saint of Maelienydd and Gwerthrynion where his churches include Nantmel, Llanbister and formerly Rhaeadr, as well as Llangoedmor and Llangynllo CRD. The last occurs as *Langwynllo* 1409, *Llangynllo* 1493.

Llanifan, Llanevan SO 1561

(hafotry) Llan Jen 1383-4, (tenement) *Llan Jevan, -Ieuan 1550,* (farm) *Llanevan 1564,* (land) *Jnestone 1568*

'John's church', with W *llan* as above and *Ieuan, Ifan* but there is little proof of any church here apart from *Bron'r Eglwys* 1833 (*Fron-yr-eglwys* 1820-1) applied to *Llanevan Hill. Llan* then could simply mean 'enclosure', its original sense; Llanifan could even be the place where *Ieuan* ap Llywelyn Fychan of Crugeryr built a house 'under Radnor Forest' (*i dan glut*) c.1460, mentioned by Lewys Glyn Cothi. See Radnor Forest.

Llansanffraid Cwmteuddwr see **Cwmteuddwr**

Llansanffraid-yn-Elfael SO 1054

Lansanfret in Elevain c.1200, Lannsanffered 1276, ymhlwyf san ffret 1447-89, Llansanphriad in Elvel 1664, Bridechurch 1291, (church of) *St Bridget in Elwael Uchmynydd 1434*

'Church of St Bride in Elfael', with W *llan, Sanffraid* and the n. of the *cantref.*

Llansteffan, Llanstephan SO 1242

Llanstyffan 1513, Llanstephan 1578, Llanystyffan wen 15thC, *Seint Stevyns parisshe 1553-5*

'Church of St Stephen', with W *llan* and *Ystyffan,* Mod W *Steffan.*

Llan-wen Hill SO 2969

Llanwen Hill 1833

In 1815 it is *Farrington Bank* (see Farrington). Named after *Llanwen* farm (*Llanwen* 1815) but without earlier sps. it is too hazardous to suggest a meaning since *Lan-* and *Llan-* in RAD are sometimes modifications of W *llwyn* 'grove' and *nant* 'stream' as well as *glan* 'bank' and *llan.*

Llan-ŷr, Llanyre SO 0462

Thlanhur 1304, ll.llyr yn Ros c.1566, Llanyr yn rose 1575-6, Llanyere 1566, Llanerry 1670, Llaneere 1753

W *llan* and the saint's n. *Llŷr* – said to be a god of the sea (evidently with a poor sense of navigation!). The first sp. above may be a manuscript error for *Thlanllur* but more early sps. are needed. The original form may have been *Llan-llŷr* (as *Llan-llŷr* CRD) but the middle *-ll-* seems to have been assimilated in the same way as *Llaneirwg* (= St Mellon's MON) dedicated to *Lleirwg* or *Lleurwg* and *Llanychaearn* CRD dedicated to *Llwchaearn*. The final *-e* of the OS map form *Llanyre* is evidently silent (despite the 1670 sp.) and has confused newcomers to the area. Some antiquarians have construed the pn. as *Llan-hir* 'long church' but that is very unlikely – only one sp. lends any support – but popular interpretation may have played a part in the development of the pn. The full n. has *-yn-rhos* 'in the moor' – not the n. of a district but a qualifier like *mynydd* in Llanbadarn Fynydd and *helygen* in Llanfihangel Helygen. The odd thing is that such qualifiers were generally used to distinguish places with identical first ns., such as *Llanddewi* or *Llanfihangel* but the only similar pn. to *Llan-ŷr* is *Llanllyr* CRD.

Llowes SO 1941

Podum liuhess in eluail 12thC, *Locheis* late 12thC, *Lewas, Lockwas* c.1291, *LLEWES* 1497, *Langhes* (= *Laughes) 1520-1, Llochaes 1517, Llowis 1513*

Gerald of Wales (late 12thC) spelt the name *Lotheis* (twice) and *Locheis* and the scholar Egerton Phillimore suggested that he may have written *Loƀeis* (representing *Loweis*) in his drafts which was mistranscribed. Sps. generally have either *-w-*, *-ch-* or *-ckw-* which are reminiscent of W *lloches, llochwes* 'refuge, place of safety' and 'room, chamber' and instances are known where *chw, w, wh* and *gw* alternate as, for example, in *gwarae* for *chwarae* 'to play' and in the rn. *Chwefri* BRE which occurs as

Hweuery 1278, *Cwmwheuery* 1299, *Wheffrey R.* c.1755, *Weuery R.* 1695, *Whefri* 1828, *Chwefru* 1833.

W *lloches, llochwes* might well have been a suitable explanation of the pn. were it not for the fact that the church is dedicated to saints *Llowes* and *Meilig* and this is confirmed by the mention of *podum* (= 'religious settlement') *Liuhesi et Lann meilic halyguessin* the Book of Llandaf 12thC, probably referring to a grant c.680. In the Life of St Gildas *Lyuhes* was said to have been joined by *Maelog* (or *Tyfaelog*) or *Meilig* in the district of Elfael. *Meilig* gave his n. to the house *Crossvilog* (SO 181446), properly *Croesfeilig* (so spelt c.1817); his reputed cross is now in the church and is almost certainly *croes Tyfaelog* in a poem of Lewys Glyn Cothi (1447-89).

Llynwene SO 1557

Llanwene 1593, Llanwenney 1666, Leinwenny 1582, Llynwenny 1610-11, Llin Wennen early 17thC, *Thleuwenny 1336*

Ffransis Payne (CrF) quotes a line of Morgan Elfael (*floruit* 1528-41), a local poet, *'Llynwene'n llawn o wenwyn'* ('Llynwene is full of poison, *or* worry'). Despite the sps. with *Llan-* (and *Lhuynweney* late 17thC) we may actually have *llynwynau* a pl. of W *llynwyn* 'standing pool'.

River Lugg : afonLlugwy

Lucge c.1025, 1085, Lugge 1231-3, 1256, Lug 1292, Luggo 15thC, *Lugg 1568, llugwy 1447-89, Lhygvvy 1572*

Identical in meaning to *Llugwy*, a tributary of the r. Conwy CRN and *Lligwy* ANG deriving from the OW root which has produced ModW *llug* 'light'. The W form has *-wy* as in *Cleirwy* (see Cleiro) and *Edwy* (see Aberedw) often abbreviated to *-o* or *-w*. *Glanllugwy* (Bleddfa) is *Glan lyggow* 1763 and is evidently pronounced 'Lanliggo'; this has W *glan* 'bank'.

Maelienydd

am glawr maelenyt c.1160, o uaelyenyt c.1190, Melenith c.1200, Maelienid 1273-4, o uelienyd 1447-89, Meleneth 1536

'Land of Maelien', a pers.n. deriving from British *Maglogenos* 'son (or descendant) of the hero', with the territorial ending *-ydd* found in, for example, *Meirionnydd*. The poet Lewys Glyn Cothi refers to it familiarly as *Gwlad Vael* 'land of Mael' where *Mael* is a short form for *Maelien*. See also *Nantmel*. Maelienydd was a *cantref* or hundred covering most of north and east RAD.

Maes-gwyn SO 9970

Maes Gwynn 1688, Maesgwyn 1670, Masegwyn 1544

'White open-field', with W *maes* which in ModW usually just means 'a field' and 'outside' (*tu maes*). The n. of a former township in Nantmel parish, now preserved in the farm n. (SN 990701).

Maesllwch SO 1740

(manor) *Maisloughe, -lowghe 1584, Mas Llwch c.1700, Maeslough House c.1817*

Maes as in Maes-gwyn and *llwch* 'lake, pool'. There are several pools in the r. Wye here. *Maes* also occurs in nearby Maesyronnen and both are likely to refer to open areas near the r.

Maesyronnen SO 1741

Maes yr Onnen 1614-15, Maesyronnen 1669, Maes-yr-onnen 1778

W *maes* as above, def.art. and *onnen* 'ash-tree'. Best known as the n. of an old Baptist chapel established in 1696.

River Marteg
Blaen marchdeg 1447-89, Marteigry 1688, Afon Marteg 1833

'Very fair' according to Melville Richards, containing W *march* and *teg*; *march* can mean not only 'horse, stallion' but something 'large, coarse' as in plant-ns. in the same way as 'horse' in E *horse-radish*. There is a good example of this in *Marchwiel* near Wrexham DNB which has *march* and *gwiail* 'rods, reeds'. *Farteg* (the maps have *Varteg*) near Pont-y-pŵl MON and two places with the same n. occur in Cilybebyll and Margam GLA and are likely to mean the same as *Marteg*.

Medwaledd SO 1583
medwaledd 1579, Mudwallet 1671, Bedwaledh c.1700, Medalweth 1670

Possibly a pers.n. *Medwal* or *Meddwal* and an ending *-edd* often used to form adjectives. Some sps. may have been influenced by W *bedw* 'birch-trees'.

Meistirrhosllawddu SO 1067
Meysterstohey 1670, Maestu Rhos-lowrddu 1688, (township) *Mystyrrhoes llowdy 1851,* (townships) *Maysters and Rhoes Lowrthi, Rhoes Lowrthy 1635, Rhoes y Rhynford alias y-Maesdir 1762*

The ns. of two small townships combined. The first may be W *mystwyr (mwstwr)* thought by Prof. G.O. Pierce to derive from Vulgar Latin *mon'sterium* (Latin *monasterium*), possibly confused with W *meistir* 'open country, plain' – but our sps. are very late. The second may be W *rhos* 'moor' with a pers.n. *Llywri*. W *-r-* and *-dd-* can sometimes interchange to produce *Llywddi*; in this case the *-r-* has been retained to produce *Llywrddi*, hence the sps. with *-lowrddu, Lowrthi* etc. The grid reference given above applies to *Rhosllawddu*; *Meistir* was near Tŷ-du (SO 1266).

Michaelchurch-on-Arrow : Llanfihangel Dyffryn Arwy SO 2450

Seint Michel c.1270, Mihelescruch alias Mihellescherch 1309, Mychaell Church 1613, ll.V'el dyffryn arw 1590-1, Llanihangle: Arro 1610, (parish) *Arrewe 1333*

The E n. is largely self-explanatory and probably derives from the W n., i.e. the common pn. element *llan* and *Mihangel* as in Llanfihangel Helygen with W *dyffryn* 'valley' and the rn. *Arwy* or *Arrow*. This occurs as *Arewe, Arue* 1256, *Hareye* 1272, *Arroy, -e* 1592, *arwy* 1447-89 and *Erge* 958; it may mean 'silvery river', containing the root *arg-* found in W *arian* 'money, silver', with *-wy* as in Cleiro (from *Cleirwy*). The modern E n. *Arrow* seems to come directly from W *Arwy, Arw* but the spelling *Erge* in an OE source suggests there was probably an older E form of the rn. (it might have produced *Erry* or something very similar) containing the same OW root *arg-*.

Milebrook SO 3172

Mile Brook 1833, Milebrook, - House 1816

The brook which joins the r. Teme here marked the parish boundary between Knighton and Brampton Bryan HER (which included Stanage RAD) and is a little over one mile from the town of Knighton.

Moel Hywel SO 0071

Moel Hywel 1820-1, 1833

W *moel* meaning 'a bare hill' and the common W pers.n. *Hywel,* often anglicised as *Howel(l)*. The hill is notably bare.

Mynachdy, Monaughty SO 2369

Manachty 16thC, Monaughtye 1578, Nonaught 1553, Mynachty 1574

'Monk's house' literally, with W *mynach* and *tŷ* 'house', as in Bugeildy, referring originally to a grange of the abbey of Cwm-hir. The n. is now borne by a large house marked *Mynaughty*

Llugw 1833; *Llugw* or *Llugwy* is the river Lugg, distinguishing this place from another former grange *Monaughty Poeth* (SO 2574) near Knucklas. Both ns. are horribly mispronounced as 'Mun'awtee' by E speakers.

Nannerth SN 9471

Nanneyrth 1184, (manor) *Nanerthgwy 1585, Maen Nannerch Goy 12thC (1426)*

W *nant* and perhaps *eirth* 'bears' though there may have been confusion with *nerth* 'strength, power'. There are several farms here distinguished as *Nannerth y llyn* 1739, *Nannerth Gwy* (see Wye), *Nannerthycha* and *Nannerth Yssa* 1697; the first has W *llyn* 'lake, pool' and the last two have *uchaf* 'upper' and *isaf* 'lower'. There was even a farm called *Nanerth dinewys* 1739: was it 'wealthy' (W *dinewyn*)?

Nant-glas SN 9965

Nant-glas 1833

'Blue stream', with W *nant* 'brook' and sometimes 'valley' and *glas* (as in Glasgwm). A small stream flows through the hamlet southwards towards Nant Treflyn.

Nantmel SO 0366

Nan Mayl 1304, Nantmayl 1259, Nanmaell 1391, Nantmel 1513

There is no *mêl* 'honey' in *Nantmel,* simply *nant* meaning 'valley' (rather than 'brook' as in Nant-glas) and the pers.n. *Mael* found also in Maelienydd and Elfael. The lack of stress on *-mael* led to shortening of the vowels to produce *-mel*. In 1371-2 this place is called *Lannail,* possibly another example of confusion of *nant* and *llan*.

Newbridge : Pontnewydd SO 0158

(bridge) *Pont Newydd 1679, antient bridge . . . called Pont newidd ar Wye, New bridge uppon Wye 1682, New Bridge, New Bridg House c.1755*

The road-signs ignore the W n. yet *Newbridge* may be a translation of *Pontnewydd* which has W *pont* 'bridge' and *newydd*. The present bridge is indeed 'new' but there was a 'new' bridge here in 1679 and an 'antient' bridge in 1682! Speed shows a bridge here on his map of the county 1610, perhaps the timber bridge mentioned by Leland in the 1530s. The village grew up around the junction of the Builth-Rhaeadr road and the road leading to the bridge over the Wye, possibly during the 18thC (a Baptist chapel was built here in 1760 and a Wesleyan chapel in 1811), and may have been known as *Pentref-newydd* ('new village'), according to Joshua Thomas 1778 and the Religious Census 1851. See also Wye: Gwy.

Newcastle SO 2663

Niwecastel 1304, Newcastell 1424-5, the Newe Castell 1545, Newcastell in Radnorsland c.1553

Once 'new' now the old castle at *Castle Ring* (SO 266636)

Newchurch : Llannewydd SO 2150

NEWCHURCH IN ELWELL 1497, Newchurch 1542, ll.newydd c.1566

Self-explanatory; there does not seem to be any record of when the church was established but its dedication to Mary may favour the 13th or 14thC (when many churches acquired a dedication to her) or a little later. The parish lay in Elfael.

New Radnor : Maesyfed SO 2160

Radnoure Nova 1277, New Radnor 1487, maes hewed c.991 (1286), Maes Hyueyd 1231, Radnor, in Welsh Maesyuet 1584

Radnor, or course, was also the n. of the lordship: *Radenovere* 1231, *Radenore, de Valle de Radenoure* 1293, *Radnor otherwise*

Radnorslande 1553-8. The n. applied first to Old Radnor and the distinguishing adjectives *New* and *Old* seem to occur from 1253 but there has been some dispute among historians with regard to the date of the establishment of the separate castle and settlement at *New Radnor* which logically must be later than that of *Old Radnor*. The castle at New Radnor, however, may be late 11thC and it is likely that many early mentions of Radnor castle refer to New Radnor as the location of the main castle of the lordship. The W n. *Maesyfed* is really no help since the manuscripts which mention it are no older than the late 13thC though they often drew on much older records. One W chronicle actually refers to *Radynor* in 1196. The W n. nonetheless appears to be very old since a poem c.1230-60 mentions *bro hyueid* (*bro Hyfaidd* in ModW) meaning 'country (or lowland) of Hyfaidd'. The same pers.n. is found in Brynhyfedd. In all likelihood, New Radnor was probably founded in an area which already bore the W n. *Maesyfed* 'open land, plain of Hyfaidd'. Nothing certain is known of Hyfaidd.

New Radnor gave its n. to *Radnorshire (Sir Faesyfed)*, the county established in 1536-42. See also Radnor Forest and Old Radnor.

Norton : Nortyn SO 3067

Nortune 1086, Nordton' 1193, Norton c.1210-18, Norton c.1291, nortyn c.1566, C(astell) Nortyn late 16thC

'North farm, north settlement', with OE *nord* and *tūn*, perhaps in contrast to Clatretune. The W form with *-tyn* is similar to *Gwestyn* (Weston) and other pns. such as *Prestatyn FLI, Selattyn* SHR and *Melltun* (Mellington) MTG

Offa's Dyke : Clawdd Offa earthwork

Offedich 1184, 1233, Offæ's dyke 16thC, Clawdd Offa early 13thC

'Earthwork of King Offa', named from the king of Mercia (757-795), thought to have ordered its construction to mark the boundary of his kingdom with Wales. W *clawdd* means the same as OE *dîc*.

Old Radnor : Pencraig SO 2559

Raddrenove 1086, Radenouram 1188, Radenor 1196 (c.1286), de Vet(er)i Rad(enore) 1293, Old Radenouere 1252, Pen Craig 1447-89, pen Kraic c.1566

Apparently, 'red hill', with OE *rēad* and *ofer* 'hill, ridge' or perhaps *ōfer* 'bank, river-bank' confused with *ōra* having a similar meaning. The sense 'hill' is likeliest because Old Radnor stands on the slopes of Old Radnor Hill and because the W n. *Pencraig* also implies the same. The els. are clearly W *pen* and *craig* 'crag, rock' and it may mean 'chief rock' rather than 'rock top'. John Leland (1536-9) applies the n. *Pencrage* to the actual hill. The n. *Maesyfed Hen (Maesuved hen* 1720) seems to be a form invented by antiquarians, adding W *hen* 'old' to the W n. of New Radnor.

Painscastle : Castell-paen SO 1646

castellum Paen 1198, Paynscastell 1542, Maud's Castle called Paynescastell 1397, a chastell Paen 1196, CASTELL PAYNE 1535 Pains Castle c.1817, Matildis castrum 1195, Castle Maude 1298

Said to have been named after *Pain* fitz John (died 1137) who witnessed a grant by Bernard de Neufmarché, conqueror of Brycheiniog (Brecknock) 1121. He may also have given his n. to *Painswick* GLO. His wife *Maud* (*Matilda* in Latin) de St Valéry gave the castle its alternative n. The W n. has *castell* 'castle' and an adapted spelling of *Pain.*

Pant-y-dŵr SO 9874

Panty Dŵr c.1755, Pant-y-dŵr 1833

A scatter of houses in 'the watery hollow' between the river Marteg and Nant Tawelan on the B. road running from Rhaeadr to Llanidloes MTG: W *pant,* def.art. and *dwfr, dŵr* 'water'.

Pennant SO 2176

Peynand, Pennadh 1332, Pennant 1392-3, y Pennant c.1700

'Stream (or valley) top', with W *pen* 'head, end, top' and *nant* as in Nant-glas and Nantmel. This is the upper part of Bugeildy parish. Wernygeufron farm (SO 224766) is called *Pennant Hall* 1833.

Pentre SO 2466

Pentre 1833

Perhaps W *pentre(f)* in its modern sense of 'village' since it stands near to the hamlet of Casgob.

Pen-y-bont SO 1164

Penpont Rhyd y cleifion 1778, Pen y Bont 1720, Pen-y-bont 1833, Pont Rhyd y Cleifion 1831

The village stands at the eastern end of the bridge over the r. Ieithon, hence its n. meaning 'end of the bridge', with *pen* as in Pennant, def.art. and *pont* 'bridge'. The lost addition *Rhydycleifion* means 'ford of the sick people', with *rhyd*, def.art. and *cleifion* which can also mean 'lepers', hence one writer has suggested that there may once have been a lepers' hospital here. It is very curious how often *cleifion* and *clafdy* 'hospital' occur in W pns.: note *Rhydyclafdy* CRM and *Pontycleifion* MER. Why are they so frequently associated with bridges and fords? Dolau Jenkin farm on the west side of the r. is *Dole Jenkyn alias lloyn y Clyvyon* 1603 and *Lloyn y Clyvyon* 1714 (with W *llwyn* 'grove, bush') and there was also a nearby *Cae Cleifion* (W *cae* 'field, enclosure') and a *Dolycleifion* (W *dol* 'meadow') opposite the house Shady Grove.

This must be the place described by Lewis Morris c.1757 as *Pontrhydycleifion* where Welsh was still spoken.

Penygarreg Reservoir see **Elan**

Perwydd SO 1566

Perweth 1662, Berwydd 1833, Perwydd 1820-1, Trefberneth 1297, Treberveth 1720

Not W *perwydd* 'pear-trees' but 'middle township, middle settlement', with *perfedd*. Some sps. show the longer n. with W *tref* as in Treddegwm and Trefonnen. *Treberfedd* is also the W n. of *Middletown* near Welshpool MTG which occurs as *Trefperuet* 1265, *Trif Bereved* 1289 and *Trefberveth under Breydin* 1290; it stands near the hill called *Breiddin*.

Pilleth : Pylallai SO 2568

Pelelei 1086, Pulele 1142, Pullelit c.1270, Pululith 1297, Pullyth, -ith 1316, Pilleth (in Bryttysh Pylale) 1573, Pylaley, Pylalley 1597

OE *pul, pyll* and *hlið* meaning 'pool slope'. Some sps. show loss of the final – (= -*th*) perhaps through confusion with the very common element *lēah* usually found as *-ley* in E pns. meaning, among other things, 'glade'. The W form seems to be *Pylallai* or *Pylale* deriving from the OE forms *Pelelei* and *Pulele*. W practice shifted the stress to the middle *-e-* (usually lost in E) and so preserved it as in the case of *Prestatyn* FLI (which would probably have developed into *Preston* if it had not been altered by W speakers).

Pool Hill hill SO 1775

POOL HILL 1820-1, Pool Hill 1833

There is a small pool here feeding a tributary of the r. Lugg.

Presteigne : Llanandras SO 3164

Prestehemed 1137-9, Presthemed c.1291, Prestmede 1249, Prestende 1378-9, Prestene 1548, Prestain A. L.Andreas B. 1573-80, Llanandras 1262, ll.andras c.1566, Llanandrewe, -andre 1536-9

OE *prēost* 'priest' and either *hæmed* 'household' – meaning perhaps 'priest's house' –or more likely OE *hemm* 'boundary' and *mæd* 'meadow'. The same word or words are found in *Upper* and *Lower Kinsham* about 3 miles east of Presteigne in

HER. Both these places and Presteigne lie near the Lugg. The W n. is a little more straightforward, i.e. 'church of St Andrew', referring to the church's dedication to that saint. In 1304 Presteigne had two fairs yearly, one on the feast of St Andrew (30 November).

Radnor Forest : Clud (Fforest Glud) hills
(wood) *Clud, which is called the chace of Radenore 1304, Forrest of Cluid 1536-9, forest of Radnor, otherwise fforest Gledey 1550, y fforest Glud 1588, Radenore Forest 1383-4*

Radnor Forest was a hunting preserve within the lordship of *Radnor* (see New Radnor and Old Radnor). The W n. *Clud* survives in *Gludre* farm (SO 1966) at the foot of Llethr Du – perhaps compounded with W *bre* 'hill' – and was the epithet borne by Einion Clud, a ruler of Maelienydd, killed in 1176. The W poet Cynddelw in his 'Marwnad' [lament] to Cadwallon ap Meurig (c.1170-79) refers to Cadwallon as 'celebrated owner' of *clud ac aeron* (*ac aeron* may be a miscopied spelling for *Cymaron*). *Clud* seems to mean much the same as W *cludfan* 'high pile' and *cludair* 'heap'.

Red Hill hill SO 1550
Red Hill c.1817, 1833

'Red' perhaps from its vegetation.

Red Lion Hill hill SO 0577
Banc-y-dolau 1820-1, Bank y ddolau 1833

The hill is named from the farm *Red Lion* (SO 066772) – so spelt 1820-1 and 1833 – an old tavern on a short road running from Llananno to David's Well (R. Colyer, *Roads and Trackways of Wales*, p.102). The older n. of the hill has W *banc* 'bank, hill', def.art. and the farm-n. *Dolau* (SO 053762) called *Dol-y-Gleision* 1820-1, an error for *Dolau-gleision* meaning 'blue, green meadows'.

Rhaeadr, Rhayader SN 9768

(castle) *Raidergoe 1191, Readr' 1339, Raydr' 1337-8, Rayader 1295, Rayadyr Gwy 1177, Rayadr ar wy c.1566, Rhayadr 1588*

'Waterfall' though there are now only rapids here in the Wye *(Gwy),* a result it is said of the cutting of a water-channel in the river during the erection of a bridge in 1780. The bard Dafydd y Coed c.1380) refers to 'Raeadyr oercri kadyr ar crwkedeu – bach bychin geinyauc wertheu' ('Falls of mighty wailing with your little pails full, little pennyworths': Glyn Williams). Local authorities and other official bodies still cling to the fossilized sp. *Rhayader.*

Rhiwlen, Rhulen SO 1349

Ruylnynnan 1232, Ryulan c.1291, Riwlen c.1566, Rulen 1559, 1670, Rewlen 1544

Possibly W *rhiw* 'hill, slope' and *glan* 'bank' which describes its position on the stream banks very well. The lack of stress on *-lan* could have 'obscured' its vowel to produce sps. such as *-len.* There is a hill *Esgair-rhiwlan* (SN 8871) in Cwmteuddwr spelt *Esgair-rhiwlen* 1833 and a *Rhulan* farm (SN 9336) in Llanfihangel Nant Bran BRE spelt *Rhiwlan* c.1817; the first has W *esgair* 'ridge'.

Rhiw'r-huad SN 9772

Rutharth 1232, Rhiwriad 1647, Roworydd 1666, Rhiwowrad 1734, Wenriad 1670, Worriade 1544

Sps. vary so much that it is difficult to be certain. It may have W *rhiw* (as in Rhiwlen), and *gwyriad* 'a bending or curving', perhaps a twist in the course of the r. Marteg, i.e. *Rhiw-wyriad* 'slope at the r. bend', confused with *rhiw,* def. art. *y(r)* and *huad, hued* 'hound'.

Rhos-goch SO 1847

Rose goch 1666, Rhosgoch 1791

'Red moor', with W *rhos* and *coch, goch* 'red' perhaps being the colour of *Rhos-goch Common* (*Rhos Goch* c.1817).

Rhos-y-meirch SO 2769

Rhose y Myrch 1791, Rhose y Merch 1728, Rhos y Meirch late 16thC

W *rhos* as in Rhos-goch, def.art. and *meirch*, the plural of *march* 'horse, stallion'. It originated as a squatter settlement on common land between Knighton and Whitton.

Rhwng Gwy a Hafren

Rwng Gwy a Hafrenn c.1500, rwng gwy a hafren 13thC, inter Vagam et Sabrinam, inter Waiam et Sabrinam 1188.

'(Land) between Wye and Severn' a loose description for those commotes and cantrefi, viz. Ceri MTG, Gwerthrynion, Maelienydd, Elfael RAD, and Buellt (the area around Builth) BRE, between those rs. There are similar ns. elsewhere in Wales such as *Rhyndwyclydach* near Swansea GLA between the Upper and Lower Clydach rs.

Rhydieithon see Llanfihangel Rhydieithon

Rhyslyn

Reeslyn 1335-6, Reslyn 1340-1, RIESLYN 1372-3, Rislyn 1612-3, Ryslyn 1360

Possibly W *rhys* 'wild' and either *glyn* 'valley' or *llyn* 'lake', perhaps *Llyn Gwyn* (SO 0164). Rhyslyn was a manor which included Gwastedyn Fawr (see Gwastedyn) and Faenor in 17thC lists; it may also have included Cornel near the Vulcan Arms (SN 9863), *Llangoze* (unidentified) and Cefn Nantmel.

Saint Harmon SN 9872

(church, *Sancti Germani 1188, Sancto Germano c.1291, san harmon 1447-89, saint harmon c.1566, Saynt Hermand 1542, ll.harmon c.1545-64.*

It is sometimes supposed that *Saint Harmon* is an anglicised sp. of *Llanarmon,* the usual way of recording the n. of a church dedicated to St Garmon, but out of a sample of 23 sps. only 3 call it *Llanarmon* and all are 16thC and later. *Saint Harmon* is actually an authentic W pn.. Typically, *Saint* (or *San[t])* precedes the saint's n. when the saint is non-Welsh or where the parish is in an area where E influence has been strong. *Saint* before the saint's n. should nearly always be regarded as a borrowing from E *saint. Saint Harmon,* however, incorporates the n. of a W saint and lies in an area which until recently was remote from E influence. What seems to have happened is that W *san(t)* was prefixed to the n. *Garmon* as a mark of honour (compare *Sanffraid* in Llansanffraid-yn-Elfael) and the *G-* of *Garmon* was dropped in the regular fashion to produce *Santarmon* and then an *-h-* intruded before the stressed vowel *-a-* to produce *Santharmon, Sant Harmon.* Finally, the sp. of *Sant* was influenced by the W word *saint* borrowed from E.

Salford SO 2362

Saltforde 1304, Saltford 1833, Salford 1487

OE *salt* 'salt' and *ford* usually meaning 'a ford' but it also had the sense 'road' as shown in *The Longford* SHR, a Roman road, and in the W borrowing *ffordd* 'road, way'; 'salt road' would make better sense here applying possibly to the road running from Knighton to New Radnor. Salt was an essential commodity in the Middle Ages (as today) and a 'salt road' was usually an important routeway.

Slough SO 3063

Sloughe 1566, Slough 1833, the Slough 1790

ME *slough* 'slough, quagmire' – the n. of a hamlet in the bottom of the valley of Clatter Brook (for which see Clatretune). *Slwch* (SO 0628) near Brecon seems to mean the same thing and occurs as *Sloghe* 1488, *Slough* or *Slozthe* 1551.

Stanage SO 3371

Stanege 1086, Stanegge 1252, Stanage 1340, Stanadge 1584, Stanedg, Stonedge 1670

'Stone edge, stone scarp', with OE *stān* and *ecg*. Stanage is dominated by an abrupt, wooded hill Park Bank Wood, once part of Stanage Park, overlooking the river Teme. Lower Stanage is by the river. *Stanage* HER and *Stonedge* DRB mean the same.

Stanner SO 2658

(way under) *Stanner 1563, Stanner Rocks 1804, a green serpentine rock called Stanner 1814*

The sps. are too late for a confident explanation but the n. seems to have originally applied to the steep rocks here and the n. is likely to have OE *stān* as in Stanage.

Storling Bank hill SO 2067

(house or inn) *Stall Inn 1815, Starling, - Bank 1833*

Apparently named from three houses on the north slopes of the hill. Without earlier spellings it is too hazardous to suggest a likely meaning: *Stall Inn* could be an attempt to make sense of *Storling* or some similar n., e.g. *stallion* – as in *Stalling Busk* YN (DEPN) and perhaps *Stalling Down* GLA – or it may be what it implies, i.e. a *stall* 'stable, stall' and *inn*. The house lies near the old road running from Bleddfa to Casgob, a former drovers' route.

Swydd a'r Graig SO 1365, 1262

South & Graig 1670, Swydd, Graig 1851, Swyd and Graige 1799

W *swydd* perhaps in the sense of 'sub-division of a *cantref* or hundred' but Swydd may be identical with *Swydd Meliniog* which is *Melynnog* 1304, (mill of) *Veyleynok* 1332, (mill of) *Soithmoleniok* 1392-3 *Melinog, Mellinog* 1722. This was not a geographical district of Maelienydd like *Swydd Dinieithon* (see Cefnllys) but a *swydd* in the sense of 'a group of tenants owing suit (or services) to the manorial mill', with *melin* 'mill' and the adjectival ending *-iog*. The 1392-3 reference says that the 'mill' lay within Swydd Dinieithon. Whatever the case may be, *Swydd* became attached to the farm of that n. (SO 130645). Within Swydd township lie *Rhos Swydd* (mountain called *Rhose y swydd* 1603., *Rhoes y Swydd* c.1755) and the hill Coed-swydd (above).

Graig appears to have been a separate township amalgamated with Swydd before 1670 and was centred on several farms south-west of Llandegle; two farms called *Graig* and one called *Graig-uchaf* (W *uchaf* 'upper') occur here in 1833. All evidently take their ns. from the stony crest of *Llandegley Rocks* (*The Rocks* 1833) since W *craig* means 'rock, crag' (as in Cregrina). The first letter of *craig* has probably changed to *G-* because of the effect of a lost def.art. W fem. nouns with an initial *b, d, g, p, t* or *c* (like *craig)* change by soft mutation to *f, dd, -, b, d,* or *g* respectively.

Swydd Buddugre SO 1069

Buthigfrc (= -fre) 1232, Soyth Uggore 1332, Soithbuthugre 1372, Soythudugr' 1360, Comm. Soededugre 1536-9

A sub-division of Maelienydd (see Cefnllys and Swydd a'r Graig) taking its n. from *Beddugre Hill (Beddygre* 1820-1, *Bedd Ugre* 1833) near Llanddewi Ystradenni. The poet Prydydd y Moch (1173-1220) mentions *uuddugre lys* (*Fuddugre Lys* in ModW) in an ode addressed to Gruffudd ap Cynan, a ruler of Gwynedd (d.1137). This may be our *Buddugre* though there are two places with this n. in DNB and in MER. All have W *buddug* and *bre* meaning roughly 'victory hill, ? victorious hill'. The poet's *lys* is W *llys* 'court, mansion'. The modern map spelling *Beddugre* is probably taken from the popular notion that the pn. contains W *bedd* 'grave' and the n. of a giant *Ugre. Swydd Buddugre* itself was contracted by a process of contraction called 'haplology' to *Swyddugre* in which similar sounding syllables – in this case *-ydd* and *Budd* - are reduced to one sound.

Swydd Dinieithon see **Cefnllys**

Swydd Meliniog see **Swydd a'r Graig**

Swydd Rhiw'r-allt

Soyth Ruyll 1332, Soithroill 1392-3, Southruralthe 1563, Soytheruyllare 1360, Comm. Soedriwalt 1536-9

The sps. are contradictory; the first element is W *swydd* as in *Swydd Dinieithon* (see Cefnllys) but the rest is difficult. Many later sps. favour W *rhiw* 'slope, hillside', def.art. and *allt* 'slope, wooded slope' but earlier spellings lend little support. Some seem to suggest W *rhwyll* 'lattice' but I do not know of any other pns. with this word. It is possible for W words ending in *-yll* to acquire a final *-t*, i.e. *rhwyll* might appear as *rhwyllt* (or some similar sp.).

Rhiw'r-allt has been identified with *Rhiwlas* near Bleddfa as well as *Old Hall* and *Weston Hall* (on the site of Upper Weston – see Weston). The actual *swydd* or 'sub-division' covered eastern Maelienydd, extending from Knighton to Llanfihangel Rhydieithon, and included part of Llanddewi Ystradenni in 1649.

River Teme : Afon Tefeidiad

Temede 757-775, Temde, Teme 1536-9, Tavidiot c.1485, taueidiat 1447-89, Dyffryn Tyveidioc 1233

Teme may derive from British *Tamētio-* and *Tefeidiad* from *Tamētiatis,* once thought to have the meaning 'dark' as in the case of *Taf* and *Tawe* but, as Ekwall remarked, darkness is 'not a very marked characteristic' of any of these rs. *Tamĕtio-* would probably have produced OW *Temeid,* MW *Tefeid* but I have not seen any clear instances of these. The W spellings with *-oc* may represent an additional form *Tefeidiog.*

The minor lordship of *Tempseter* (OE for 'settlers by the river Teme') in SHR on the north side of this r. was called *Cwmwd Dyffryn Teveydiat* in a W source c.1540-50; this has W *cwmwd* 'commote, division of a *cantref* or hundred', *dyffryn* 'valley' and the rn.

Traean-y-mynech

Taianymenighe (=Traian-) 1578, Trayan y menighe 1594

'The monks' third', with W *traean,* def.art. and *mynech, myneich,* a plural of *mynach* 'monk'. This was the area of Cwmteuddwr parish formerly held by the abbey of Strata Florida CRD.

Trawsgoed SO 0359

Trouscoed 1688, Trowscoed 1822, Troscoed 1670

Loosely, 'area covered with trees', with W *traws* 'across' and *coed* 'trees, wood'. A former township of Llan-ŷr parish, still partly wooded.

Treddegwm

No sps. seen; presumably W *tref* 'township' and *degwm* 'tithe, tenth part'. A former township of Llanbadarn Fawr parish. The n. seems odd since most townships paid tithe to the church.

Trefonnen SO 0662

Treve onen 1562, Trevonnen 1650

W *tref* as in Treddegwm and *onnen* 'ash-tree'. The ns. of trees are often found in combination with *tref* in Wales; the sense may be elliptical, i.e. indicating an area characterised by a particular species of tree. This area is now part of Llandrindod.

Tre-goed SO 0555

tref y koed c.1562, trekoyd 1649, Trecoyd 1670

Trecoed is the form given by OS maps and many other sources but one would expect *Tre-goed,* with a soft mutation after *tref,* a fem. noun. Without earlier sps. it is difficult to be certain but the c.1562 spelling suggests that a def.art. *y* has been lost: the pn. may have been *Tre'r-coed* (with the 'affixed' form *'r* of the def.art.). That would explain the *-c-* in modern sources. The meaning would be 'township associated with the wood' or 'township characterised by trees or a wood'. There are similar examples of pns., especially in south Wales, losing their def.arts., e.g. *Trecastell* (or *Trecastle*) BRE deriving from *Tre'rcastell* (with W *castell* 'castle').

Trelowgoed SO 0863

Trelowgoed-isaf, -uchaf 1820-1, Trellowgoed 1833

Without earlier sps. it is difficult to make convincing sense of this. Presumably it has W *tref* like Trefonnen and Tregoed and perhaps *gloyw* 'bright' and *coed* 'wood, trees' – perhaps a species of tree?

Trewern SO 1243

Trewern 1251, Trawern 1254, Trefwern 1513

W *tref* and *gwern* probably meaning 'township characterised by wet land' since *gwern* indicates a place favoured by alder trees, a wet area, as well as the tree species itself. It is a common house and pn. throughout Wales. *Trewern* MTG near Welshpool, for example, had an E equivalent and occurs as *Alretone* 1086, *Orleton* 1255, *Trewern alias Olarton* 1558-90. *Trewern* (SO 2259), Llanfihangel Nant Melan, is *Treweryn* 1340-1, *Trefwern* 1360-2, *Trewerne* 1582.

Trumau SN 8667

Trumau 1833

W *trumau*, pl. of *trum*, meaning 'ridges, summits'. The OS map reference confines the n. to a marshy moorland near the spot height (1709 ft.) but the OS map 1833 applies it to a broader area extending westward to Hirnant Claerwen (stream), much more appropriate to its meaning.

Tynllan SO 1462

Tynllan 1851

No earlier sps. seen but it is likely to be W *tyddyn, tyn,* and the common *llan* referring to the church of Llandegle, and meaning 'church dwelling, - cottage', a little odd since it applies to the central township of the parish; perhaps it replaced *Tre'rllan* or *Trellan* 'church township, the township with the church in it'. There is another *Tynllan* (SO 1858) in Llanfihangel Nant Melan referring to a farm.

Uwchcoed

Huccoyd 1338, Ughcoyd 1360-2, Ughcoid 1424-5

'Above wood', with *uwch* 'above' and *coed* as in Iscoed which was the lower part of Gwerthrynion in contrast to Uwchcoed.

Walton SO 2559

Walton 1304, 1408, Walltown 1371-2

There are many similar ns. in England once often interpreted as 'settlement of the Welsh' or something much the same but there are other possible explanations with OE *w(e)all* 'wall' and *wælle, wiell* 'stream, well'. Walton is more than a mile from Offa's Dyke and I do not think that *Walton* can refer to this earthwork; there is a small stream passing through the village.

Waun-wen SO 2154

Waynwen, Waynewen 1685, Wenwaine, . . .nwayn 1670, Waun-wen 1833

W *gwaun* 'moor' and *gwen,* the fem. form of *gwyn* 'white, fair'; the loss of *g-* in *gwaun* is likely to be a result of the effect of a lost def.art. by mutation (compare Tre-goed) or a preposition. Westwards there is a hill *Carreg Gwyn* 'white rock' and both ns. could refer to the paleness of the rocks here.

Wegnall SO 3263

(Philip of) *Wygenhale 1339, Wigenhale 1527, Wegnall 1566*

OE meaning 'Wicga's nook', a pers.n. and *halh. Wiggenhall* NRF means the same.

Weston: Gwestyn SO 2170

Westune 1086, Weston 1329, 1699, ynhir gwestyn 1447-89

'Western settlement or farm', with OE *west* and *tūn,* surviving as the ns. of the farms *Upper* and *Lower Weston,* both on the west side of the r. Lugg or Llugwy. The W form cited above means 'in the land of Weston' (*yn* 'in' and *tir* in the form of a nasal mutation) and is similar to the W n. for *Weston Madoc* near Montgomery: *yg gwestun* 12thC and *Gwestyne Vaddok* 1559 (with a pers.n. *Madog*).

Weythel SO 2357

Waythell 1550, (brook) *Wadele 1536-9, Gwaythell 1558, Waytholl 1670,* (brook) *Weyhill, Weyhill Mill 1828, Blayney waythell 1713*

The n. of a former township; the stream is now called *Gilwern Brook.* More sps. are needed but Dr Gelling has suggested ME *waite* 'watch; look-out place' and *hill* but which 'hill' is meant? Is it *Weythel Common* immediately south of the hamlet? Since it was also the n. of a stream it could have W *gwaedd* 'shout', meaning a 'roaring' stream; it is worth comparing the r. Ieithon.

Whitton: Chwityn SO 2767

Whytton 1332, Whitton 1304, Whittyn 1513, ll.ddewi yn hwytyn c.1566

OE or ME meaning 'Hwita's farm' according to B.G. Charles but 'white settlement' is equally possible. Were the houses whitewashed? The W form is indicated by the 1513 and c.1566 sps. above, with *Chw-* (often pronounced like E *Wh-*) replacing OE *Hw-* or ME *Wh-* and *-tyn* replacing *-tūn, -ton* as in Weston: Gwestyn. The church is dedicated to David (Dewi) but *Llanddewi* does not appear to have been the common n. for the village for W speakers despite the c.1566 sp.

Wolfpits SO 2158

Wolfpitt 1548, 1552, Woolpits 1624, Wolfpits 1828

What it suggests, 'a pit for trapping wolves'. There are two places called *Woolpit* in SFK and SUR with the same meaning. Note also Bleddfa.

Womaston SO 2660

Wymoneston 1304, Wymaston 1343, Womaston 1543, Woomaston 1694

OE meaning 'Wigmund's settlement'. Domesday Book's *Ulfemestune* has been wrongly identified with this place.

River Wye: Afon Gwy

Guoy c.800, Gui, Guy c.1150, gwy c.1300, (inter) Vagam 1188, Waie 1204-14, Waye late 12thC, Owy, caullid in Englisch Wy 1536-9

Various suggestions have been made including W *gwyro* 'wind, bend', and 'running water' from British **Uogio-* or **Uogia-* but it may be a pre-Celtic n.

Yardro SO 2258

Kay Lloyd or the Yardro 1715, the Yardro 1749, Yardro Farm 1771, Artraw 1828

Uncertain; a similar n. *le Yardroue* occurs in CHE 1325 and is thought to mean 'rough ground near an enclosure', with OE *geard* (which gives us ModE *yard*) and *rūh* 'rough ground'. The variant n. 1715 stands for ModW *Cae-llwyd,* 'grey (or brown) enclosure or field', but it may not be a genuine W n. for Yardro since it occurs in a late deed; during the 17th and 18thC we often find spurious 'alternative' ns. for places, partly because of lawyers and their scribes misunderstanding earlier deeds and partly because lawyers were over-scrupulous in recording what they took to be genuine variant ns.